Bankruptcy to Billions

'The book provides a first hand insider's account of the remarkable transformation of Indian Railways. It contains valuable insights on alternative strategies which were able to get results when conventional prescriptions could not be applied.'

—**Montek Singh Ahluwalia**, Deputy Chairman, Planning Commission, Government of India

'The role of infrastructure in development has long been recognized, and its efficient management has become even more pertinent today. This book is a splendid account of how India addressed successfully the key management problems in the important infrastructure sector, Indian Railways, under dynamic leadership. It should be required reading for policymakers worldwide.'

—**Jagdish Bhagwati**, University Professor, Economics and Law, Columbia University

'The authors have demonstrated that smart and honest bureaucrats can bring efficiency in the most difficult turnaround situation. Using the powerful leadership attributes of creative thinking and strategic action, they have demonstrated that it is possible to blend commercial objectives with social obligations.'

—**N.R. Narayana Murthy**, Chairman of the Board and Chief Mentor, Infosys Technologies Limited

'This book provides valuable insights and knowledge that would be inspiring and useful for a wide variety of readers and particularly those who are engaged in the transformation of organizations currently facing difficulties, but striving to attain financial success, and greater relevance to society.'

—**R.K. Pachauri**, Director-General, The Energy and Resources Institute and Chair, Intergovernmental Panel on Climate Change

'Can hundred-year-old bureaucracies change? Can politicians combine political and commercial savvy? Can politicians, bureaucrats and technocrats collaborate and innovate? The transformation of Indian Railways provides a compelling answer to these questions. By creatively combining deep political sensitivity with astute commercial and operational excellence, Indian Railways has demonstrated that profitable transformation of public service is possible and can be done in a very short period of time. *Bankruptcy to Billions* offers a well-documented and compelling insight of a unique journey. A must read for all interested in the management of large enterprises—public or private.'

—**C.K. Prahalad**, Paul and Ruth McCracken Distinguished University Professor, Ross School of Business, University of Michigan

Bankruptcy to Billions

How the Indian Railways Transformed

Sudhir Kumar
Shagun Mehrotra

with a Foreword by
MANMOHAN SINGH, Prime Minister of India

OXFORD
UNIVERSITY PRESS

OXFORD

UNIVERSITY PRESS

YMCA Library Building, Jai Singh Road, New Delhi 110 001

Oxford University Press is a department of the University of Oxford. It furthers
the University's objective of excellence in research, scholarship, and education
by publishing worldwide in

Oxford New York

Auckland Cape Town Dar es Salaam Hong Kong Karachi Kuala Lumpur
Madrid Melbourne Mexico City Nairobi New Delhi Shanghai Taipei Toronto

With offices in
Argentina Austria Brazil Chile Czech Republic France Greece Guatemala
Hungary Italy Japan Poland Portugal Singapore South Korea Switzerland
Thailand Turkey Ukraine Vietnam

Oxford is a registered trademark of Oxford University Press
in the UK and in certain other countries

Published in India
by Oxford University Press, New Delhi

ISBN 13: 978-0-19-806085-7
ISBN 10: 0-19-806085-8

Typeset in Adobe Jenson Pro 11.5/14
by Excellent Laser Typesetters, Pitampura, Delhi 110 034
Printed at Parangat Offset, New Delhi 110 020
Published by Oxford University Press
YMCA Library Building, Jai Singh Road, New Delhi 110 001

To our inspiration
Guddi and Nina

Contents

Contents

Figures and Tables

Figures and
Tables

FIGURES

TABLES

सत्यमेव जयते

प्रधान मंत्री
Prime Minister

Foreword

Foreword

Since the onset of liberalization in the early 1990s, the Indian economy has experienced sustained and rapid growth. The quest of the UPA Government has been to enhance the rate of growth and also ensure that its benefits reach the lowest levels of the economic pyramid. The success of the Indian Railways during this period epitomizes the achievements of the UPA Government in meeting the dual challenges of reducing extreme poverty while sustaining economic growth. This book on the transformation of the Indian Railways describes how progress is possible while benefiting, rather than burdening, the poor.

By leveraging existing assets, essentially by operating 'faster, longer, and heavier trains', the Indian Railways have dramatically enhanced capacity and productivity. The experience of the Railways demonstrates that optimizing on existing assets is as critical as seeking incremental investment to meet future needs. Through dynamic and differential pricing that is customer centric, the Railways have become market oriented and yielded enormous economic as well as social gains. These reforms required questioning past beliefs, practices, and associated assumptions. This book explains why the success of the Railways is not just a spillover of economic growth, but achieved through fundamental improvements in the very functioning of the organization.

I congratulate the authors for this publication. I hope this work inspires further innovation in infrastructure reforms that are vitally required for sustained economic growth and poverty reduction.

New Delhi
3 December 2008

(Manmohan Singh)

Acknowledgements

Acknowledgements

First and foremost, the authors would like to thank Lalu Prasad, Minister of Railways, who led this exemplary process of inclusive reforms. This book has greatly benefited from the Minister's support. Not only did he provide unrestricted access to information and analysis, but also provided his valuable time and deep insight into the reform strategy. Further, the core team of officers and staff working in the Office of the Minister provided invaluable support, in particular R.K. Mahajan, K.P. Yadav, Bhola Yadav, and Binod Kumar Srivastava, who was extremely generous to share his office for several months during a critical stage of the book writing.

Equally important has been the support of the 1.4 million dedicated and committed Railways staff—the champions of this transformation. This book is a reminder of the tireless efforts of the Railways family, as it continues to connect markets and people across India. In particular, the authors would like to thank members of the Railway Board: K.C. Jena, Chairman, Railway Board; Sowmya Raghavan, Financial Commissioner; Shri Prakash, Member Traffic; S.K. Vij, Member Engineering; R.K. Rao, Member Mechanical; Sukhbir Singh, Member Electrical; and S.S. Khurana, Member Staff. Further, thanks are due to M.S. Gujral, R.R. Jaruhar, R.R. Bhandari, R. Sivadasan, L.R. Thapar, Sumant Chak, Radhu Dayal, and C.B.K. Singh, for sharing their expertise and unique knowledge that has greatly benefited this book. S.B. Ghosh-Dastidar, Ranjan Jain, P.P. Sharma, Manish Tewari, K.P. Yadav, and Ivan Crowley are few we name, without reducing the debt we owe to others, as they carefully read drafts, gave very useful comments, and corrected many misconceptions. Further, Anjali Goyal, Sanjeev Jain, G.D. Paul, and Neeraj Gupta have been a perennial source of data that needed numerous verifications, resulting in reliable and extremely detailed statistics and information on accounts, although the authors retain any responsibility for errors. K.C.N. Menon, Dalip Singh, S.K. Roy, Ranmat Singh, and Anuj Prasad, in addition to patiently and persistently enduring the needs of this extremely demanding book writing, carefully vetted the list of references, figures, and other data. They have truly worked way beyond the call of duty, allowing us to meet what seemed an impossible deadline. S.N. Mishra and Rajendra Kumar took great personal care to support the authors during the preparation of this book, once again enduring extremely demanding schedules and long work hours.

Moreover, we would like to thank Columbia University for allowing space, time, and resources to this work. At the university, many debts are owed for invaluable direct and indirect insight, criticism, help, and advice: Bob Beauregard, Jagdish Bhagwati, Stephan Brown, Sumila Gulyani, Peter Marcuse, Richard Markwald, Vijay Modi, Arvind Panagaria, Henry Pinkham, Cynthia Rosenzweig, Smita Srinivas, Jeff Sachs, Elliott Sclar, Joe Stiglitz, and Ambassador Frank Wisner; colleagues at the World Bank—Anand Rajaram, Antonio Estache, Ede Ijjasz, and Wendy Ayres—and Tony Gomez-Ibanez at the Kennedy School. Special thanks are due to Petros Mavroidis, for being a wonderful friend and mentor, and Hans Smit for all his words of wisdom.

At Oxford University Press, we would like to thank the editorial team for making this book a reality despite the unusually short timeframe, while maintaining their high standards of rigour and quality of publications. Furthermore, feedback from three anonymous referees is gratefully acknowledged.

Finally, our deepest gratitude to our families and friends, who allowed us to focus solely on work at the cost of spending time with them.

December 2008

Sudhir Kumar
New Delhi

Shagun Mehrotra
New York

Preface

Preface

The Indian Railways is a unique state-owned enterprise because of its size, ownership structure, and one-hundred-and-fifty-year-old history. These attributes, among others, make it a complex, intriguing, and thus fascinating subject to understand. The Indian Railways is one of the world's largest state-owned enterprises, a utility under a single management, second only to China's. With around 1.4 million employees and 1.1 million pensioners, one of the world's largest railway networks—over 63,000 kilometres of routes—running approximately 13,000 trains each day, including 9000 passenger trains, the railways is a Ministry within the Government of India. Indian Railways carry over two million tons of freight and some 17 million passengers between 7000 railway stations each day. A fleet of 200,000 wagons, 40,000 coaches, and 8000 locomotives achieves this. To fathom the scale, consider the fact that each day Indian trains travel four times the distance to the moon and back.

Like many state-owned infrastructure service providers, the Indian Railways was over-manned and charging subsidized fares, as it uneasily attempted to balance between conflicting commercial and social objectives. In the financial year 2000–01, the Railways defaulted on dividend payments to its primary investor, the Government of India. The cash balance shrunk to US$ 83 million (Rs 359 crore) and its operating ratio—a key measure of efficiency (operating expenditure over operating revenue)—peaked to 98 per cent. While nominal improvements were made in the following years until 2004, the financial condition remained precarious.

However, in the next four years, under a populist political mandate that did not allow conventional policy prescriptions—in the words of the Railways Minister, 'no privatization, no retrenchment, and no fare-hike'—the finances of the Indian Railways, as well as the quality of service provided, transformed. The operating ratio of the Railways improved to 78 per cent, thereby generating a cash surplus of US$ 6 billion (Rs 25,000 crore) with astounding growth in freight volumes, as well as market share and earnings. However, what make this transformation unique are the distinct approaches and their swift accomplishment. Precluding the pursuit of textbook solutions, namely privatization, retrenchment, fare-hikes, and the like, the financial health of the Railways was restored without burdening the millions of poor Indian travellers and employees. The transformation of the Indian Railways is attributed to a management strategy that optimized the functioning of this complex institution on its apolitical variables. The core supply side strategy can be summarized in three words, each worth more than a billion dollars in annual surplus: faster, longer, and heavier trains. Likewise, the gist of the demand-side strategy is dynamic and differential pricing that is customer-centric. This book unpacks each of these, as well as several other detailed strategies of the transformation.

The first chapter, 'Bankruptcy to Billions!', provides a snapshot of the transformation, and focuses on the management strategy adopted to implement change. The second chapter, 'Political Economy of Reforms', captures the fine balancing act between commercial and social obligations of the Railways, including the political mandate of 'no privatization, no retrenchment, no fare-hikes' and how creatively the space for reforms within this mandate was identified.

The third chapter, 'The Market', captures the analysis that allowed for the development of a socially responsible and commercially viable understanding of specific business segments that offered opportunities for transforming the Railways without burdening the poor sections among the consumers.

The fourth and fifth chapters, 'Milking the Cow' and 'Service with a Smile', articulate the 'how' aspect of the reform strategy. While the former captures the supply-side strategy, the latter articulates the initiatives that led to market-oriented reforms in the freight, passenger, and other coaching segments.

Finally, 'Outcomes, Sustainability, and Replication' summarizes not just the financial gains, but gains in productivity, political capital, and a host of other vectors. Further, challenges to sustaining the present generation of reforms as well as scope for replication are explored for other public-sector-dominated services like water supply, electricity, irrigation, and public transportation.

Thus, the purpose of this book is to understand three associated aspects. First, to briefly define the existing policy approach to reforming large state-owned enterprises like the Indian Railways, and to assess associated limitations. Second, to explain in-depth, why the prescriptive approach to reforms was not applicable to the Indian Railways, and how the railways transformed from near bankruptcy to a US$ 6 billion annual surplus through an egalitarian approach. Third, to distil lessons for other state-owned enterprises that provide infrastructure, with the aim of developing an understanding towards an unconventional approach to inclusive reforms.

Rather than arguing what the Railways ought to be in an ideal world, this book presents a pragmatic view of how the railways transformed in a short period of four years focusing on what works. The book demonstrates that a profitable railways is a prerequisite to reconciling social and commercial obligations of the Railways as in an expanding pie scenario resulting from productivity gains, the Railways is able to deliver its social obligations without threatening the viability of the organization. This reorientation of the Railways from a 'profit averse' to a 'profit oriented' organization was an essential prerequisite to galvanize the energy of the Railways towards the common goal of inclusive transformation.

However, this transformation would have been impossible without the extraordinarily talented and committed pool of Railways staff who are equally good at planning mega-projects as they are at translating them into reality. The Railways has faced several challenges in the past—at the time of its introduction, the task of post-Independence nationalization and unification, the operational crisis of the 1980s, and, most recently, the financial crisis—and on each of these occasions the institution has emerged stronger, and this transformation is yet another example of the same. The success of the Railways is the result of the hard work put in by its talented employees, rather than the result of chance, or just a booming economy. It results from the combination of excellent people in the Railways and some timely and smart strategies.

On each occasion, the leadership of the Railways played a catalytic role in galvanizing its internal strengths. In the most recent transformation, a constructive politico-bureaucratic engagement was a lynchpin for introducing change. Routine engagement between the political leadership and the bureaucracy fostered mutual trust, regard for mutual differences, and an environment that allowed for creativity, innovation, and change. Integrity and commitment of the top leadership were essential ingredients for fostering this trust and commitment to constant change. The present achievements of the Railways are just a glimpse of its magnificent past. In the words of the Minister, '... this is only the beginning; the true potential of the Railways is yet to be unravelled.'

1 Bankruptcy to Billions!

This is an overview of how Indian Railways was transformed in four years (2004–08), counter-intuitively, under populist leader Lalu Prasad, current Minister of Railways. What makes this surprising is that while retaining state ownership, the Railways graduated from near bankruptcy in 2001 to US$ 6 billion[1] annual cash surplus in 2008.

Lalu asks a gathering of people from his constituency, *'Tum ko malum hai ki Railway ne pichle char saal mein 70,000 crore rupaiyah munafa kamaya'* (Do you know, in the last four years the Railways earned Rs 70,000 crore in profits)?

A voice from the crowd responds, *'Na Saheb'* (No, sir).

Lalu asks, *'Char saal pehle Hathua se Siwan ka yatri kiraya kitna tha'* (Four years ago what was the passenger fare from Hathua to Siwan)?

The people respond, *'Jee saat rupaiyah'* (Sir, Rs 7).

Lalu, *'Abhi kitna hai?'* (How much is the fare now)?

The audience responds, *'chaar rupaiyah'* (Sir, Rs 4).

Lalu *'Pahle kisi rail mantri ne yaatri kiraya kam kiya tha'* (Did any Railway Minister ever reduce passenger fares in the past)?

The audience responds, *'Na Saheb'* (No, sir).

Lalu *'Hamne har saal yaatri kiraya kam kiya, phir bhi 70,000 crore kamaye. Kaisa laga?'* (Every year, I reduced the passenger fares, yet the Railways made a profit of Rs 70,000 crore over the last four years. What do you think?)

The excited audience screams back, *'Wah Wah Wah! Yeh toh kamaal ho gaya.'* (Congratulations sir, this is great).

Soon after the Railways had earned a cash surplus of Rs 15,000 crore (US$ 3.5 billion) in 2006, the Minister was keen that this financial gain translate into rewards for the Railways employees, and tangible benefits for the poor travellers that rely on the Railways for transportation. Of his propositions, the most striking was his insistence on reducing second class passenger fares by a rupee per passenger. The Railway Board members were perplexed, 'Why reduce just one rupee? In most transactions nowadays, a rupee (2½ cents) has no value. This fare reduction will cost the Railways Rs 250 crore (US$ 58 million) and the passengers will hardly benefit.' The Minister responded in typical colloquial fashion: *'Hathua ki gwalan apna dudh Dilli me nahi balki Siwan me bechati hai. Aur Hathua se Siwan ka kiraya maatr saat rupaih hai'* (a milk vendor from Hathua—the Minister's area in Bihar in eastern India—sells her milk not in Delhi, but in Siwan. And the train fare from Hathua to Siwan is just Rs 7). *'Lagta hai ki air-conditioned office me rahne walon ko yeh ehsaas nahi hota ki ek garib gwalan ke liye ek rupaih ke kya kimat hoti hai'* (it seems that those who reside in air-conditioned offices do not realize what a rupee means to a poor milkmaid). Further, the Minister elaborated that it was likely that all her relatives lived within a 70 mile radius, and thus most of her work- and life-related train trips were within this small radius. Apart from being colloquial and populist, this argument was also persuasive and at the end of this exchange the Railway Board agreed to a Re 1 fare reduction.

This anecdote is grounded in a larger reality. A trip analysis revealed that the average fare of 88 per cent of railway travellers—namely all suburban and ordinary passenger train users—is about Rs 10. As the Minister got the second class passenger fares reduced by Rs 3 over four years, the minimum passenger fare went down from Rs 7 in 2004 to Rs 4 in 2008. This is a 42 per cent reduction. As a result, a rupee reduction in fare is not just symbolic, but has a substantive effect on the total fare for these travellers. During this period, the annual bonus for railway employees was increased from 59 to 73 days of their respective wages. However, what is critical is that poor consumers, as well as railway employees, directly benefited from the Railways's financial gains—annual cash surplus grew to Rs 25,000 crore in 2008 (US$ 6 billion) and the Railways was not accused of profiteering.

Further, this proved to be a popular initiative to the extent that it was central to the Minsiter's public meetings even in his constituency, as is evident in the exchange quoted at the beginning of the chapter.

This exchange between the Minister and his constituency draws attention to the need to balance commercial objectives with social concerns such that the market metric is balanced with the societal one. Not only is there a compelling moral imperative, but also a political and economic rationale to address the needs of the marginal if the overarching commercial goals are to be achieved and sustained.

Despite the Odds

Indian Railways is a web of life that weaves people and markets into a diverse yet unified India. The railway is as critical for the poor as it is for the economy. It is the only affordable means of transport for millions of commuters, as well as aspiring migrants who travel to realize their dreams in the city. On the other hand, freight trains haul critical commodities that crank the wheels of the economy—taking raw materials to power, steel, and cement plants, as well as foodgrains to ration shops and fertilizer to farmers. This grand institution, one of the largest employers with 1.4 million employees and 1.1 million pensioners, faced a severe financial crisis in 2001 when its cash balance shrank to a paltry Rs 359 crore (US$ 83 million), operating ratio[2] deteriorated to 98 per cent, and it defaulted on payment of dividend to the Government of India. The severity of this financial crisis is aptly captured in the Mohan Committee Report (Mohan 2001b).

To put it bluntly, the Business As Usual Low Growth will rapidly drive IR (Indian Railways) to fatal bankruptcy, and in fifteen years GoI (Government of India) will be saddled with an additional financial liability of over Rs 61,000 crore (US$ 14.2 billion). ... On a pure operating level IR is in a terminal debt trap and can only be preserved by continuing and ever increasing subsidies, year-on-year, from the central government. As is well known such subsidies are not available [pp. 180–1].

Over the 1990s, the core profit-making freight segment grew at the 'business as usual low growth' rate of 2 to 3 per cent, and wages grew faster than the growth in labour productivity (Mohan 2001b: 178). As a result, in the five years that led up to the 2001 crisis, the Railways's expenses grew at over 13 per cent per annum, while its revenues lagged at 8 per cent. This was unsustainable as the Railways was unable to generate sufficient cash to even cover the cost of replacement and renewal of its aging assets. The World Bank (2006) noted that had the Railways made adequate provision for depreciation it would have been in the red.

It is to be noted that IR's operating ratio of 0.96 is substantially understated, as the provision of depreciation is well below actual requirements. If IR were to make adequate provision for annual asset renewal, a fortiori if it were to make adequate provision for the large backlog of overdue equipment and track renewals, as well as pension accruals, in normal commercial accounting term, it is very likely that IR would be a heavily-loss-making entity—in fact one well along the path toward bankruptcy, if it were not state owned [p. 5].

This under-provisioning for depreciation endangered operations and led to stacking up of replacement arrears year after year. To liquidate these arrears the Government of India had to constitute a special railway safety fund worth Rs 17,000 crore (US$ 4 billion), two-thirds of which it gave as a dividend free grant and the rest was financed through a safety surcharge on passenger fares. While improvements were made up to 2004, the Railways's financial condition remained precarious. However, in the next four years a populist political mandate did not allow conventional policy prescriptions; in the words of the Railway Minister, 'No privatization, no retrenchment, and no fare hike.' Yet, counter-intuitively, the finances of Indian Railways have been transformed.

The cash surplus of the Railways rose steadily from Rs 9,000 cr. in 2005 to Rs 14,000 cr. in 2006 to Rs 20,000 cr. in 2007. The August House would be happy to know that in 2007–08, we will create history once again by turning in a cash surplus before dividend of Rs 25,000 cr. (US$ 6 billion). Our operating ratio has also improved to 76%. Indian Railways is a government department. However, we take pride in the fact that our achievement, on the benchmark of net surplus before dividend, makes us better than most of the Fortune 500 companies in the world [Budget Speech, Minister of Railways, 26 February 2008: 1].

There has been a complete reversal from a predicted terminal debt trap to a cash-rich organization with a bank balance of over Rs 22,000 crore (US$ 4.7 billion). In 2008, the Railways internally generated six times more cash than its annual debt repayment obligation of about Rs 4000 crore, (US$ 0.93 billion), making it a grossly under-leveraged organization. This was acknowledged by global investors in the United States when they subscribed to the Indian Railway Finance Corporation's bonds over four times in a matter of a few hours, at a rate of 5.94 per cent in 2007—a coupon rate that is better than what was offered to the best private firms in India. Moreover, at 76 per cent, the Railways's operating ratio is better than the operating ratio of Chinese rail as well as class one American railroads; and its 21 per cent return on net worth is better than that of some of the blue chip Sensex companies in India. Earlier, the investible surplus was insufficient to finance the replacement of aging assets, but in 2008 the same railway generated an investible surplus of Rs 20,000 crore, and its capital expenditure tripled compared to 2001 (see Table 1.1).

Freight and passenger volumes clocked a compounded annual growth rate of 9 per cent between 2004 and 2008. During this period, asset and labour productivity grew at twice the rate of the 1990s. Earlier it was heading towards bankruptcy because expenses were growing 5 per cent faster than revenue, but now the Railways has become super solvent by reversing the pattern, where revenues are growing over 5 per cent faster than expenses (see figure in Table 1.1).

Freight business profits have boomed because of growing volumes, declining unit costs (from 61 to 54 paise), and increasing unit revenue due to selective fare hikes (from 74 to 93 paise, see

TABLE 1.1 Financial indicators		
	2001	2008
Cash surplus before dividend (Rs crore)	4790	25,006
Investible surplus (after payment of dividend) (Rs crore)	4204	19,972
Capital expenditure (Rs crore)	9395	28,680
Fund Balance (bank balance) (Rs crore)	359	22,279
Operating Ratio	98.3%	75.9%
Ratio of net revenue to capital-at-charge and investment from capital fund (return on net worth)	2.5%	20.7%
Debt service cash overage ratio	1.74	6.53

Source: Statistics and Economics Directorate, Ministry of Railways

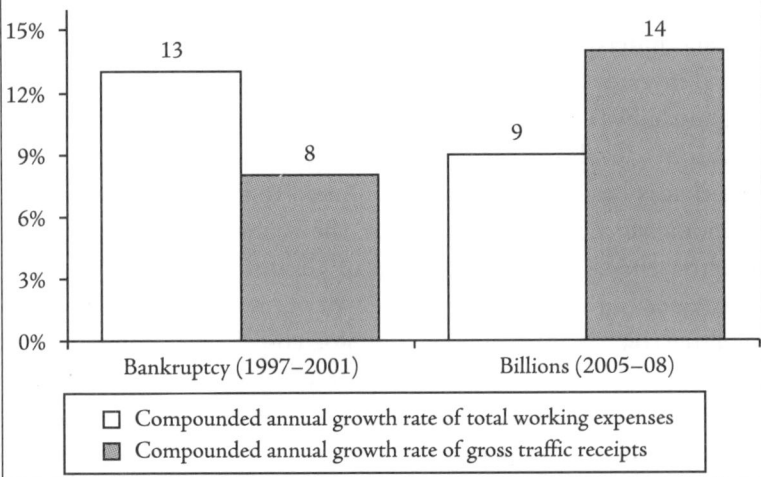

☐ Compounded annual growth rate of total working expenses
▨ Compounded annual growth rate of gross traffic receipts

Figure 1.1). Despite reduction in passenger fares of most travel classes, losses in the passenger business have been curtailed by virtue of stable unit costs (38 to 39 paise) and increase in unit revenue due to a change in product mix in favour of high-value and high-margin air-conditioned and long-distance travel segments (from 23 to 26 paise, see Figure 1.2). Further, the growth rates of 'other coaching' as well as 'sundry earnings' have doubled from around 10 per cent in 1991–2004 to over 22 per cent in 2005–08. This has been achieved by enhancing non-passenger fare income through leveraging eyeballs and footfalls of travellers and by tapping unutilized parcel capacity.

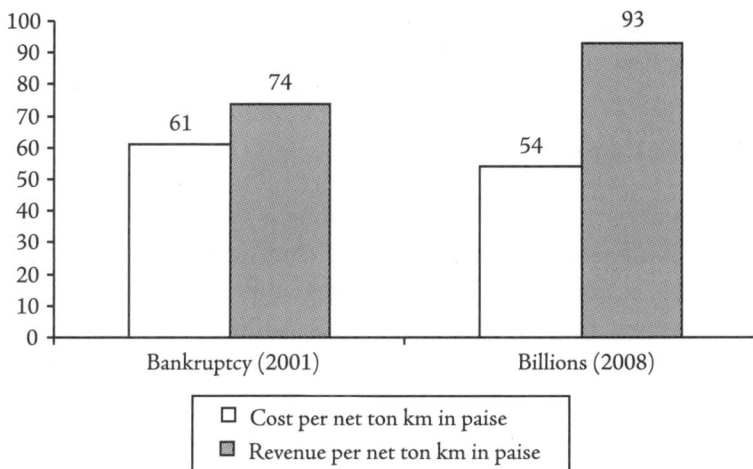

FIGURE 1.1: Freight unit revenue and cost

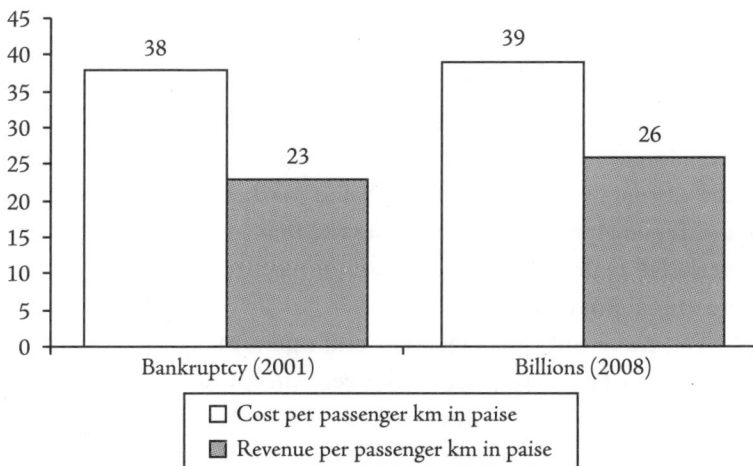

FIGURE 1.2: Passenger unit revenue and cost

Source: Statistics and Economics Directorate, Ministry of Railway, 2008.

Apart from the financial transformation, customers have benefited from faster, safer, and better services. The identity of the organization and morale of employees have received a boost, as has the stature of the Minister, who was earlier ridiculed as the 'joker of Indian politics' and is now referred to as 'Professor Lalu' as he lectures the students of IIM Ahmedabad, Harvard, and Wharton, teaching how to blend populism with profitability.

MANAGEMENT STRATEGY

What makes this transformation unique is the distinct approach and its swift accomplishment. The financial health of the Railways has been restored without burdening millions of poor Indian travellers, or railway employees. This astonishing financial performance of the Railways has drawn national and international acclaim, surprised policymakers, bankers, and the people alike, and has shell-shocked some sceptics. This transformation is not merely a result of commodity cycles or a booming economy, but structural change resulting in significant gains in operational efficiency. Nor is it a result of creative accounting or unsafe overloading of wagons, instead it is largely due to labour and asset productivity gains. The core supply-side strategy can be summarized in three words, each worth over a billion dollars in surplus: faster, longer, and heavier trains. The demand-side in another three words: dynamic, differential, and market-driven. To execute these demand and supply strategies the management created cross-functional teams, leveraged existing resources, and synergized operational interventions.

Yet, these nifty management and operating strategies are not new, and in fact date back to the very inception of the Railways. In this regard, Lalu is often asked by business school students, 'If it is so simple, why wasn't it done earlier?' Lalu's response is straightforward, 'Indian Railways has a *huge* potential. It's like a Jersey cow; if you don't milk it, the cow falls sick. Therefore, we are milking the cow fully and taking good care of it.' However, what remained unanswered was why this metaphorical 'cow' was not 'milked' earlier. There may have been several other factors at play, but a conceptual cause was a populist political mandate at loggerheads with the policy prescription of experts. As captured in the Mohan Committee Report (Mohan 2001a), it was this conflict that had debilitated the Railways.

On the one hand, IR is seen by the government, and by itself as a commercial organization. It should therefore be financially self-sufficient. On the other hand, as a department of government it is seen as a social organization which must be subservient to fulfilling social needs as deemed fit by the government. It is now essential for these roles to be clarified [p. 5].

Over a 150-year history, the Railways has confronted several formidable challenges: at its inception, the Railways was built under the difficult conditions prevalent in the mid-nineteenth century, then there were the post-Independence challenges of nationalization and modernization followed by the critical operational crisis in the 1980s. On each of these occasions, the Railways's talented and technically sound staff rallied around shared objectives and strategies, and demonstrated its ability for adaptive resilience, emerging stronger each time. This is because of two core institutional strengths of the Railways—namely its people and its systems. It is staffed by some of the finest bureaucrats and technocrats recruited through extremely competitive civil service exams. And it has an equally strong organizational structure with its robust field units, as well as well-articulated procedures and processes. Yet, this formidable institution was unable to cope with the financial crisis of 2001. This was primarily due to a structural shift triggered by the liberalization of the Indian economy in the 1990s. On the one hand, the Railways's profitable freight and air-conditioned passenger segments became vulnerable to stiff competition from alternate modes. On the other hand, the shrinking fiscal space resulted in declining support from the federal coffers. To compound the crisis, the Railways continued to be burdened with social obligations like low passenger fares. Consequently, the rising cost of operations was often offset by increasing tariffs in the lucrative freight and air-conditioned passenger travel segments, further eroding the Railways's market share in these segments.

To revive the financial health of the railways, experts recommended restructuring—passenger fare hikes, retrenchment, corporatization, and regulation (Mohan 2001b). However, there was no political space to implement these recommendations as it entailed sacrificing the interests of the railway employees as well as poor travellers. This inherent contradiction between the policy prescriptions and the political mandate led to a deep-rooted cynicism within the staff (Tandon 1994). 'The feeling has built up at all levels that the solutions lie above their level: a "They" complex; only "They can decide". ...The whole attitude builds a sense of helplessness and there is growing evidence of lack of commitment and involvement, a rigidity and drift (p. 7)'.

To break free of the widespread cynicism within the Railways, the staff needed to be reassured that commercial objectives and social considerations were compatible and indeed could be reconciled. The essence of this challenge confronting the Railways is captured in Prime Minister Manmohan Singh's inaugural address to the *Economic Times*' award ceremony for corporate excellence: 'The challenge before the political leadership in India today is to meet the aspirations of an energetic new India, and, at the same time, take care of the concerns of the less endowed, less privileged sections of our society, who are no less energetic [9 October 2006: 19].

Translating the mission of inclusive reforms—defined here as meeting commercial objectives without compromising the needs and aspirations of the poor—into actions, required a deep understanding of the political economy of Indian Railways. Moreover, it required a re-conceptualization of what reforms actually meant.

Reform is, in the final analysis, about changing mindsets. We must have the courage to think out-of-the-box. We must have the courage to think anew. To question old beliefs. To seek new pathways. As an old Chinese saying goes—a road is made by walking. We must learn to walk in new directions and create new roads to progress (9 October 2006: 19).

Despite having the will to think anew, how were market considerations harmonized with societal obligations? The 'split-personality', as the Mohan Committee experts described the conflict between commercial and social objectives, implied that there were only two possible outcomes of any policy initiative—either a loss on the political objectives while gaining on the commercial front or vice versa. Initiatives like increasing lower class passenger fares or sanctioning non-remunerative new lines fall in these categories. But increasing axle load to carry more freight or adding coaches to a popular passenger train enhances earnings per train, and is welcomed by consumers as well as benefits the Railways. Thus, policy outcomes fall into four categories, not two (see Table 1.2).

Each policy initiative was examined to determine the type of outcome. The strategy was to look beyond the obvious, to reconcile perception with reality, and to find out if the conflict between the political mandate and commercial objectives was the principal reason for the deteriorating condition of the Railways. For example, the total cost function of the Railways is largely inflexible because the Railways does not have the political space to retrench at will, to shut down, or even sell loss-making branch lines and business units. On the other hand, if the total cost is distributed over greater volumes,

TABLE 1.2
Feasible set of reform outcomes

	Commercially viable	Commercially unviable
Politically desirable	Outcome 1: Win–Win Reform efforts that fall in this category create win-for-all outcomes and thus face no resistance. For example, increasing the length of a popular passenger train enhances earnings per train, and is welcomed by consumers because it helps clear long waiting lists.	Outcome 3: Exclusively social returns Reforms of this nature are extremely contentious because there is tremendous political will, yet these are commercially unviable. For example, maintaining loss-making branch railway routes (or opening new ones) for marginal communities in remote areas.
Politically undesirable	Outcome 2: Exclusively commercial returns Reforms that are socially suboptimal and commercially viable are contentious because they lack political will, typical of the 'split personality' scenario of the Railways. For example, increasing passenger fares in second class ordinary passenger trains.	Outcome 4: Lose–Lose These outcomes are neither politically desirable nor commercially feasible. For example, new railway projects sanctioned on constricted departmental considerations compromise the institutional objectives.

unit cost declines. This volume-driven strategy has no political implications and benefits customers and railways alike. Similarly, while increase in passengers fares is extremely politically sensitive, increasing yield per train by adding more coaches to popular trains benefits waitlisted customers and railways alike. A relook at the hidden opportunities behind the thick veil of the political economy led to some striking discoveries. Through a rigorous analysis it was demonstrated that about 80 per cent of the Railways's revenue streams, as well as investments, are not politically sensitive and can be market driven (see Figures 1.3 and 1.4).

As illustrated below, the entire freight, parcel, and air-conditioned passenger segments are apolitical and can be market driven. The Mohan committee attributed the declining market share, even in bulk commodities, to freight fares cross-subsidizing

22%

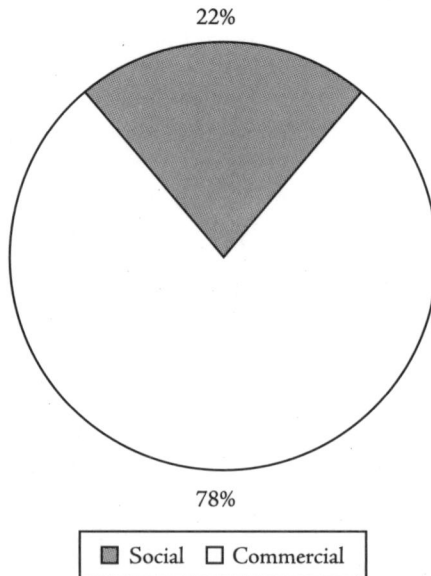

78%

□ Social   □ Commercial

FIGURE 1.3: Traffic revenue in Fiscal Year 2007–08

Note: Total revenues for fiscal year 2007–08 was Rs 71,720 crore (US$ 17 billion). The social components include all earnings from passenger trains with the exception of air-conditioned and first class. Essentially, income from poor passengers is included and from non-poor excluded.
Source: Computed with data from Ministry of Railways (2008).

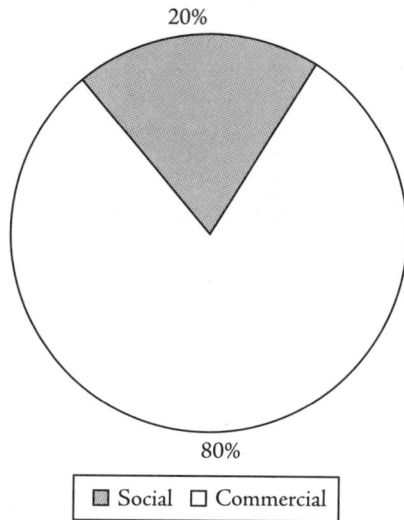

20%

80%

▨ Social ☐ Commercial

FIGURE 1.4: Investment in Fiscal Year 2007–08

Note: Total investments for fiscal year 2007–08 was Rs 28,680 crore (US$ 7 billion). The social components include all expenditure in new railway lines, urban transportation projects, and gauge conversion. Essentially, expenditure for projects considered to be pro-poor is included, and the rest is excluded. By and large these proportions have been consistent over the last few years.

passenger services. However, as these cross-subsidies affected all commodities equally, why was there a dichotomous response in the transportation of finished products versus raw materials? For instance, while the market share of finished products like steel and cement declined sharply in the 1990s, the share of iron ore, coal, and other minerals remained stable. Both iron ore and steel are heavy commodities. But there is a distinction. In the case of iron ore, railways provides a door-to-door service—from the mine-pithead to the factory. Thus, the rail freight is equal to the total 'door-to-door' logistics cost borne by the customers—implying that the incidental costs of rail transportation are negligible. On the other hand, in the case of steel, railways provide a 'station-to-station' service. But steel is neither produced nor consumed at railway stations. As a result, in addition to the rail freight, the customer incurs incremental costs due to multiple transfers, bridging, warehousing, inventory, and the like. In essence, rail freight is only a small component of the total door-to-door logistics cost to the customer in the case of steel. Truckers, on the other hand, provide door-to-door service for transportation of steel and the incidental costs of road transportation are negligible. Thus, the Railways has a strong competitive edge in transportation of iron ore, while it is extremely vulnerable in the case of steel. To compound the problem, the Railways determined freight rates based on the value of the commodity rather than elasticity of demand. Relatively low-cost raw materials like iron ore were charged less while expensive finished products like steel were charged more. Based on the competitive strength of the Railways in the marketplace, in fact, the pricing policy for these two types of commodities should have been the reverse—more for iron ore and less for steel. As a result, the Railways was losing market share in steel while retaining the same in iron ore. In sum, the differential loss in market share was not on account of cross-subsidies, but due to misconceptions about the competitiveness of the Railways in the marketplace and a monopoly mindset.

Likewise, while successive ministers were reluctant to increase non-air-conditioned class passenger fares, what prevented the Railways from decreasing air-conditioned class one and two fares to face the onslaught of low-cost airlines? Thus the Railways's core profit-making business segments were under threat largely due to an aversion to profit, a monopoly mindset despite shrinking market share, poor commercial orientation, and lack of customer focus (Mohan 2001b).

In the same vein, barring construction of new railway lines, urban transportation, and some gauge-conversion projects that are sanctioned on political considerations, the remaining 80 per cent investment decisions are not spurred by political compulsions. On the contrary, it was found that the bulk of investments were not yielding results due to narrow departmentalism that routinely compromised institutional goals for departmental gains (Tandon 1994). For instance, investments made in the 1980s and 1990s of Rs 30,000 crore (US$ 7 billion) for strengthening the railway track structure did not yield commensurate results, not because of political interference but on account of lack of cross-functional coordination and risk aversion to increasing axle load.

Such analysis was done policy by policy, business by business, and activity by activity. It revealed that space for reforms was extensive. For instance, there are no political compulsions in operation, maintenance, train examination policies, and other day-to-day operations. Conflicts between the social and commercial obligations of the Railways were not the primary cause of despair. Therefore, the focus of the reforms shifted from no privatization, no retrenchment, and no fare hike to identifying, expanding, and multiplying win–win outcomes and leveraging them to maximize financial returns without social costs. As an immediate step, care was taken to minimize lose–lose outcomes by inducing synergies between departments, establishing cross-functional teams, encouraging project-focused coordination with well-defined targets under the close eye of senior management.

While this understanding was a prerequisite, it was not sufficient. The whole organization could be galvanized by the transformational mission of inclusive reforms, only if the top leadership demonstrated fairness. Achieving this was a herculean task for Lalu because of the image associated with his rule in Bihar. Initial incidents like the Minister's in-laws forcing a change of platform for a Rajdhani Express train in Patna at the last minute or party workers found travelling in a higher class than assigned on the ticket, seemed to confirm people's apprehensions. However, on each occasion the Minster stood firm and directed railway staff not to be intimidated and to enforce rules in an impartial manner. Moreover, senior railway officers had been worried that transfers and postings decisions would be coloured by caste considerations—favouring some social sub-groups. But the Minister maintained a hands-off approach, allowing the Railway Board autonomy in making merit-based decisions regarding transfers, postings, and the awarding of contracts. This gradually fostered mutual trust and understanding between the bureaucracy and the political leadership.

It is not uncommon for the railway bureaucracy to hold politicians in contempt as they confuse the democratic mandate of 'no privatization, no retrenchment and no fare hike' with political interference, namely meddling with railways's everyday management. In this case, the bureaucracy learned to respect the political mandate and the political leadership reciprocated by not interfering with the routine operations of the Railways. Such constructive politico-bureaucratic engagement allowed for translating the mission of inclusive reforms into concrete outcomes. Further, a deep emotional connect and commitment for reform was fostered. As members of the core reform team often summarize, 'We eat, drink, sleep, and dream railways'.

Breaking the Myths

Once this critical hurdle of establishing trust and emotional connect with the staff was overcome, it unleashed the Railways's core strengths—its talented staff and robust systems—to initiate change. In the spirit of 'questioning beliefs' and 'seeking new pathways' assumptions about the nature of business and its purpose were revisited. This included variables like the very nature of the railways business and its rationale, the cost structures, revenue streams, competitive strengths, relative elasticity of price and non-price factors, its variability and sensitivity to load and length of train, as well as its manipulation.

First was the notion that the Railways is a monopoly service provider and required tariff regulation. However, this notion was inconsistent with declining market shares. In practice, the Railways was facing a competitiveness problem characterized by poor growth rates, falling market shares, and low or negative margins. The railways was losing out to alternate modes—pipelines, airlines, roadways, shipping lines, and the like. Thus a grounded view of the railway was that of a transporter operating in a competitive marketplace where it enjoyed an edge in some profitable segments and not in others. This erosion of competitiveness was unlikely to be solved by regulation; instead it required offering a superior and compelling value to the customers. Thus the focus shifted from tariff regulation to reducing unit costs, improving yields, margins, market shares, productivity, product mix, and quality of service with a customer focus.

Second, in the past profit had not been the primary focus. Instead, the Railways's operations and technology-based considerations dominated. This lack of profit orientation is not confined to public utilities like the Railways but, as observed by Carlos Ghosn, CEO of Nissan Motors, is also seen among large private corporations.

Nissan wasn't really engaged in the pursuit of profit. ...They were selling cars without knowing if they were taking losses or making profits. ... Sure, executives discussed profitability, but the company wasn't managed to that end. And when profit's not a motivating element, it won't simply materialize as a result of good luck. You have to place profit at the center of your concerns. No magic is going to bring it about [Ghosn and Ries 2005: 98].

Yet, the new-found profit orientation of the Railways differs from that of private corporations because it focuses on earning profits while serving the interests of the common people. This required a new perspective as well as business savvy: spotting, seizing, and cashing in on business opportunities were of essence. A striking example that demonstrates this savvy is a fivefold growth in freight earnings from transporting iron-ore for export—from 900 in 2004 to Rs 4400 crore in 2008 (US$ 209 million to over a billion). The international prices for iron ore soared while the cost of mining remained relatively stable, thus iron ore mining corporations began reaping windfall profits. The railways has a formidable competitive edge in transporting this commodity due to geographic conditions, bulk quantities, and long distances from mine to ports. In addition, the Railways recognized that its services were underpriced; this was reflected in a long waitlist of over 10,000 indents—requests for rakes to transport iron ore for export. Through consecutive freight rate hikes in iron ore for export, the freight rates were increased by nearly 400 per cent and the Railways cashed in on this opportunity offered by the global commodity boom, resulting in an additional Rs 9000 crore (US$ 2 billion) in profits over the last four years. Such opportunities are short-lived and therefore timely action is of essence.

Third, were assumptions regarding variability of costs. The railway's financial code prescribes the long-term variable costs of the Railways as 78.5 per cent. An analysis of the variability of costs from 1983 to 2004 revealed that while operating expenses have increased ten times at nominal (current) prices, at real (constant) prices they have decreased marginally. Meanwhile, the Railways's throughput measured in gross ton kilometres has more than doubled. This illustrates that increase in unit costs is predominantly a result of inflationary pressure and not on account of growth in throughput. Thus the variability of cost in railways is substantially less than that prescribed in the code even in the long term and is negligible in the short term. This relative insensitivity of unit cost to output is due to economies of scale, slack in the system, and improvements in operating strategy and technology. For instance, with half the number of wagons and locomotives, the railway now carries twice the amount of the load due to gains from technological improvements. Likewise, as the operating strategy changes to run heavier and longer trains, unit costs decline. This is because cost of operation is relatively insensitive to the load and length of the train as heavier, longer trains require the same crew, engine, tracks, and the like. This analysis was central to the scale-driven strategy to increase freight volumes, reduce unit cost, gain market share and margins, and make billions of dollars in profits.

Fourth is a pricing policy based on affordability. Poor passengers and low-value commodities like iron ore and minerals were charged much less than wealthier passengers and expensive commodities, usually finished products. However, ministers are not concerned whether steel freight is higher than iron-ore or the other way around. While politicians are hypersensitive to pricing for poor passengers, they also welcome fare reductions in air-conditioned segments. With this new insight, the pricing policy was creatively modified. Now the pricing for the freight, parcel, and air-conditioned passenger business segments is market driven and customer centric, while affordability-based pricing continues for low-end passenger segments. The past policy of across-the-board increase in prices to compensate for rising costs has been replaced with a policy of selective price increases based on the relative competitive strengths of the Railways. Freight charges have been increased where the Railways has a competitive edge and decreased where it is lacking. Further uniform pricing across seasons, routes, to-and-fro traffic flow, succeeded in the planned economy of the statist era, but has little relevance in a liberal economy. It has therefore been substituted with a differential and dynamic pricing policy. Substantial discounts are offered during the lean season and loading in empty returning trains. On the other hand, surcharges are levied during the busy season and on congested routes.

Fifth, the past fixation with price per passenger or per ton has conceded ground to yield per train, unit costs, margins, product mix, and the like. The profitability of a train is a function of several variables including price and non-price variables like occupancy rates, carrying capacity, load and length per train, and aspects that determine asset utilization. In turn, each of these variables is further a function of other variables, like carrying capacity depends on axle load, design of the wagons and coaches, tare weight, volumetric capacity, and density of the commodity. Thus the focus has now shifted from pricing per passenger or per ton to maximizing profits through yield and margins per train.

Sixth, emphasis on construction and procurement of new assets has been replaced with a focus on asset maintenance, enhancing productivity, and better utilization—namely by operating faster, longer, and heavier trains. Reducing the seven-day turnaround[3] time to five days enabled the Railways to run an additional 230 trains each day on average. All else being constant, incremental revenue from just these trains amounted to Rs 10,000 crore (US$ 2.3 billion). Next, by adding an extra 6 tons of load per wagon, the Railways transported 90 million tons of incremental load each year or Rs 6000 crore (US$ 1 billion) in incremental revenue. Further, by attaching 3000 additional coaches in popular mail and express trains with long waiting lists the Railways earned an incremental revenue of Rs 3000 crore (US$ 0.7 billion).

Finally, the Railways has traditionally been an insular organization driven by its processes and products. But now the focus is on value creation and customer satisfaction and a tech-savvy approach. This requires an agile and outward-oriented management strategy. Information technology and strategic alliances have been leveraged to create value and improve the quality of service so as to provide cheaper, safer, and more reliable travel. For instance, e-payments, value-added services such as half train load as opposed to a full train load of 2500 tons, multiple location unloading facilities, and faster delivery of cargos have benefited the freight customers as well as the Railways. Likewise, travellers have benefited from systemic changes in services like e-ticketing, nationwide train enquiry call centres, as well as better catering, cleanliness, and improved ambience of stations and trains. In essence, with this new perspective towards the nature and rationale of the railway business, the management confronted the formidable task of execution such that it translates to improved profits.

CRAFTING THE COALITION OF THE WILLING

Post 2001, under the uncertainty of the corporatization proposed by experts, the morale of the railway staff was low—apprehension, anxiety, and fear were the order of the day. There were concerns about job losses and pension panic was on the rise. To instil confidence within the bureaucracy and provide a sense of mission to the staff, the Minister's stance and phraseology were of essence. The political mandate of no privatization, no retrenchment, no fare hikes was reinvigorating. The Minister constantly referred to the Railways as a *sone ki chidia*, a golden bird with great potential, and this switch from critique to compliments was a first step in reinstating confidence among the railway employees. In essence, the political mandate not only reflected the needs of the electorate, but also provided security and restored confidence within the bureaucracy.

Second, to empower the railway staff the Minister adopted a hands-off approach to day-to-day management of the Railways—

including finalizing of tenders, evaluation of bids, award of contracts, and staff transfers and postings. Instead, he focused on finding qualified and talented officers for the job. This was in the same spirit as the routine corporate practice espoused by Jack Welch: 'I have no idea how to produce a good [television] program and just as little about how to build an engine. ... But I do know who the boss of NBC is. It is my job to choose the best people and to provide them with dollars' (Slater 2003: 17). Welch goes on to say that he gets rid of these people if they do not deliver (Pandya and Shell 2005: xvii). In contrast, the approach adopted by the Railways's leadership was to stand by people if they failed to deliver despite their sincere efforts. This was not just a constraint of the bureaucracy—where retrenchment is not a viable proposition—but was also a matter of principle. Gradually, as the coalition of reformers was extended across the organization, authority for decision making was decentralized, for example now the discretion to grant discounts on incremental freight is with staff at field units. Customers who approached the Railway Board with their grievances in the past, now resolve most concerns at zonal levels. Devolution of powers not only empowered the staff, but also motivated the organization to act.

Third was working with consensus. While such deliberative decision making was frustrating and a cause of delays, it was worthwhile because it developed a deep ownership for change. For instance, increasing axle load or improving train examination practices required inputs from various experts with allegiances in different departments—mechanical, financial, traffic, electric, civil—and participatory decision making required resolving these conflicting views, yielding robust solutions. On occasions, consensus required compromises to accommodate conflicting views. For instance, when the reformers proposed to introduce free upgrading of passengers to fill vacant seats from lower to upper classes, the Finance department was resistant, insisting that it would lead to losses. After a lot of debate and delays, a compromise solution was finally arrived at whereby the upgrading scheme was tried on a few trains for only a few days. Once the pilot was a success, the scheme was scaled up nationwide, and the new Finance Commissioner championed it. Finally, once consensus-based decisions were arrived at, scepticism in implementation was not tolerated. In essence, decision making was consultative because it not only yielded better solutions but also instilled a deep-rooted ownership within the management resulting in quick implementation.

Fourth was incentivizing performance within the bureaucracy. Unlike the private sector, where degrees of performance are rewarded with differential pay, perks, and career trajectories, in the Railways, quite like other public sector enterprises, promotions are seniority based and salaries and perks are uniform across comparable grades, irrespective of individual performance. Further, the organizational culture of the Railways was one of a large family where quality of work and sharing of benefits were valued more than performance-based discrimination. Thus it is not uncommon to find railway employees whose several generations have worked for the organization—in some cases up to five consecutive generations have worked exclusively with Indian Railways. In response to this organizational culture, a strategy of leveraging the deep-rooted commitment and loyalty that the staff has towards the organization was adopted. To motivate this 1.4 million strong bureaucracy a combination of normative incentives was adopted, differentiated across various staff grades—from frontline workers to senior management. For instance, on a field visit the Minister met some frontline workers, namely gangmen and key men, whose working conditions were dire—barehanded and poorly clad in severe winter, working with rather basic tools. In appreciation of the critical role of 200,000 such frontline workers, the management allocated uniforms, gloves, and better tools. Likewise, to terminate the practice of train crew carrying dry meals to cook en route, the railways now provides subsidized meals in railway canteens to drivers and the rest of the train crew on duty. Similarly, for senior officers, the perks were revised to include a house help, laptop computer, mobile phone, personal car, and opportunity for short training abroad. Such perks are common in the Indian corporate sector that has traditionally paid less than its multinational counterparts. While these are small gestures and were required more for improving the working conditions of the staff than motivating them, it clearly demonstrated that the institution cared about the employees. Further, to reward performance when the Railways achieved its mission, 600 million tons and a cash surplus of Rs 10,000 crore (US$ 2.3 billion), senior management approved group cash awards for the teams that had contributed to the success. In essence, through a combination of normative and economic incentives the reformers motivated railway employees to seek ownership of the change and improve their productivity, while non-performers were not particularly penalized.

FROM IDEAS TO ACTION

While re-conceptualizing the Railways's business to reflect its ground realities and motivating a team committed to reform were essential steps of the management strategy, next was implementation. Swift execution was central to success. Operating longer and heavier trains had been debated for decades within the Railways but management could not take decisions that would result in implementation. The Railways's narrow departmentalism, monopoly mindset, and cynicism towards the political mandate had been obstacles in the past. However, with a new mindset, this business perspective was translated into action through five critical management interventions. Indian Railways graduated from just lumbering along to becoming a winner by deploying a combination of management strategies: setting stretched targets, leveraging resources to optimize existing assets, working through cross-functional teams, fostering alliances, investing strategically, adopting a deliberative and calibrated approach, and chasing projects to swift completion to reap high returns.

First, thinking beyond the resource constraint required leveraging resources such that aspirations exceeded the resource endowment of the Railways. Here innovation and asset optimization—as opposed to asset accumulation—were central. The strategy was to fully utilize assets by running faster, longer, and heavier trains.

Second, was coordination and cooperation requiring functional and spatial synergy as well as complementarities among various kinds of policy interventions, such that the sum of parts was much greater than the whole. This was achieved through establishing cross-functional task forces that were assigned specific decision-making tasks to be delivered in a time-bound manner. Third was to fill in the gaps with strategic investments adopting a systems-based approach to improve the utilization of existing assets—low hanging fruits. Low cost, short gestation, high return, and rapid payback were the criteria for these investments. Such investments included lengthening platforms to accommodate longer trains and ameliorating network bottlenecks like small segments of weak railway tracks in high-density networks. These interventions were accorded top priority and authority was devolved to allow swift implementation. Fourth, strategic alliances were forged to meet soaring demand, co-opt competition in areas where the Railways lacked competitiveness, and forge long-term alliances with existing customers so as to offer better service. Fifth was a deliberative and calibrated approach where projects were first piloted to learn, revise, and scale up in a phased manner. A classic example of this incremental approach was the gradual increase of axle load in small increments of 2 tons on select routes and a few trains, gradually spreading across the high-density network. However, presiding over these themes was the organizational mission to champion inclusive reforms—the political mandate of transforming the financial condition of Indian Railways without burdening poor travellers and railway employees.

To implement the strategy the management had to aggressively chase targets. The chase had several elements. First was setting stretched targets. Second was to pursue the set of strategic inputs simultaneously. Third, timing, because in this context when to act is of essence. Fourth, change was induced by demand. Finally, in contrast to the tedious process of decision making, where the Railways had in the past deliberated over critical issues like increasing axle load or introducing more efficient train examination practices, when it came to implementation it was surprisingly swift because the Railways has the technical prowess, procedures, capacities, and discipline to implement both large and small changes.[4]

Sustainability and Replication

The outcome has been a win–win for the railways with over US$ 6 billion in annual profits, better service for its customers, and a positive profile for the Minister, but there are some who are sceptical of the long-term implications. Four concerns dominate in this regard. On one hand are those who believe that the Railways's performance is a result of accounting jugglery or is at the cost of safety, while others state that the successful transformation of the Railways is a result of plucking low hanging fruits or due to an upswing in commodity cycles.

These concerns for sustainability are substantially misplaced. While the Railways's freight business benefits from surging demand in a booming economy, to cash in on this demand required structural improvements in the functioning of the Railways. As for its accounts and financial statements, these are verifiable like those of any other public enterprise. Further, during the 1990s, the Railways was unable to afford replacement of overaged assets. In contrast, for the fiscal year 2008–09, assets, internal generation, and non-budgetary resources are contributing 78 per cent of an annual plan outlay of Rs 37,500 crore (US$ 8.32 billion). Likewise, all operational changes in the Railways are strictly scrutinized by its Research, Designs, and Standards Organization (RDSO) as well as an independent Commission of Railways Safety, administered by the Ministry of Civil Aviation, Government of India. Additionally, the number of train accidents has declined from 473 in 2001 to 194 in 2008, the allocations for replacement of overaged assets have increased from Rs 2300 to Rs 7000 crore over the same period, and the Railways's profits have soared—implying that safety, productivity, and profitability exhibit a complex interdependence.

As for the low hanging fruits argument, there is little substantiation. The scope to optimize utilization of railways's existing assets—through 'innovations in systems, processes, policies, and technology'—along with augmenting its capacity to respond to future demand, provide a strong base for perpetuating the present success of the Railways. This is partially demonstrated in the Railways's ability to sustain significant growth in traffic earnings for four successive years. Moreover, for a glimpse into how underutilized the Railways's existing assets are, consider the following comparisons with Chinese and American railroads. While the state-owned Chinese railroad has a comparable network and about the same number of passenger kilometres, it carries four times more freight than its Indian counterpart. Likewise, for the freight only, class one American railroads employ about one-tenth the labour force of Indian Railways but carry three times the amount of freight. Thus, the scope to continue improvements in the utilization of existing assets in Indian Railways is immense but less known.

For the short term, Indian Railways is investing much more than before in building as well as acquiring additional assets. Over the last seven years the annual budget has tripled from Rs 11,000 crore in 2003 to Rs 38,000 crore in 2009. Likewise, in the same period, the production of engines has increased from 180 to 500, wagons from 6000 to 20,000, and construction of new broad-gauge routes from 1000 to 3500 km. These investments will yield greater productivity gains because the capacity of new wagons is between 22 and 78 per cent more than the old ones. Likewise, the capacity of new passenger coaches has been enhanced by 5 to 20 per cent. To augment the capacity of the existing rolling stock, they are being retrofitted and the production of lower capacity coaches and wagons is being phased out.

For the long term, along with the introduction of capacity- and efficiency-enhancing technology, systems, and procedures, the Railways will be investing about US$ 53 billion (Rs 230,000 crore) during the next five years for enhancing capacity. All this investment is being strategically channelled to projects that have a commercial orientation. For instance, through route-wise planning, the entire high-density network's capacity will be augmented on a priority basis over the next five years at a cost of Rs 75,000 crore (US$ 17.4 billion). Likewise, priority is being accorded to strengthening iron ore and coal routes so as to carry 25 tons axle load. Finally to enhance capacity in the long term, dedicated freight corridors are being developed along the length and breadth of the country to match the golden quadrilateral and its diagonals. To meet a surge in demand, factories are being built to manufacture engines, wagons, coaches, and their parts and multi-modal logistics parks are being developed as well. In essence, Indian Railways is investing to sustain its growth trajectory.

Sustaining the management impetus after Lalu and his team leave office is less of a concern as the policy reforms have been embedded in the institutional DNA by mainstreaming systemic and procedural reforms. These have become part of the organizational routine, manuals, and to some extent norms. This can largely be attributed to leading change while respecting and strengthening the organizational identity and morale of the employees. Through a consensus-based approach, the reforms have developed deep roots within the institution.

However, what remain a real threat for the future of the Railways are three critical, yet little discussed factors. First is the importance of macroeconomic stability that was characterized by low inflation and interest rates. This is critical to reduce unit costs at current prices—a lynch-pin of the Railways's recent financial transformation. Over the 1990s, a combination of low growth in the freight business segment, at an annualized rate of 2 per cent, clubbed with high inflation, averaging 11 per cent per annum, resulted in costs increasing faster than revenues, leading to a financial crisis. However, between 2005 and 2008, the combination of macroeconomic stability characterized by annualized inflation rates of 5 per cent, a booming economy, and upswing in commodity cycles provided a fantastic opportunity for the Railways. The Railways seized this opportunity by increasing freight volumes at 9 per cent per annum for four consecutive years. This was 4 per cent more than the average rate of inflation. Consequently, nominal (current) costs declined by 2 per cent each year and the Railways's freight unit cost declined from 61 paise in 2001 to 54 paise in 2008, resulting in a doubling of profit margins despite no across-the-board increase in fares. Further, passenger losses reduced because while the cost remained stable, the product mix was changed in favour of high-value high-margin business segments. However, high inflation and high interest rates may reverse this virtuous cycle of declining unit cost, improving profit margins, and gaining market shares.

Second, unlike in the past when Indian Railways was the preferred employer for talented youth, including elite IIT and IIM graduates, there is now a gradual disinterest in working with the Railways largely due to the competition from private employers. Third, sustaining an internal drive to constantly innovate so as to create value for the customers such that the Railways is the preferred mode of transportation in various freight and travel business segments.

This successful transformation of Indian Railways consists of some transferable lessons that can be replicated in other public utilities as well as large corporations that are increasingly organized like large bureaucracies. The conventional prescriptions of corporatization, privatization, retrenchment, fare hikes, and independent regulation often work wonders in sectors where user fee is not politically contentious, as in the case of telecom and aviation. A classic case of such efforts is that of the telecom industry in India. However, there is a need to rethink this textbook approach to reforms in sectors like energy, water supply, irrigation, and railways where these are politically infeasible. Three striking lessons are outlined. First, counter-intuitively, the experience of the Railways's transformation demonstrates that commercial objectives and social obligations can be reconciled. This can be achieved by dissecting business segments into political and apolitical ones and then further disaggregating into nano constituents so as to identify an apolitical variable that can be manipulated to improve profitability without compromising the interests of the political constituencies—in the case of the Railways it was poor consumers and railway employees. In essence, an in-depth business and political analysis is a prerequisite for crafting an effective strategy.

Such analysis reveals that there is immense scope for expanding desirable win–win outcomes where the social and commercial objectives are met simultaneously. For instance, across sectors there is enormous room to improve efficiency by optimizing an underutilized system, fixing loopholes, and reducing revenue losses. To translate this insight into action requires working across departmental silos, introducing a commercial orientation to the organization, and breaking free of a monopoly mindset. Above all, this requires a productive politico-bureaucracy interface such that the bureaucracy respects the political mandate and in return the political leadership refrains from interfering with the routine functioning of the bureaucracy.

Second, thinking anew. In a fast changing external environment it is critical for public utilities to question past assumptions about the nature of the business, its cost structures, and pricing, and to respond appropriately. For instance, the widespread obsession with construction, procurement, and expenditure needs to give way to effective and efficient utilization of existing assets for enhancing productivity.

Third are the *big five* approaches to implementation: namely, setting stretched goals, cross-functional and spatial coordination, strategic investments, fostering alliances, deploying a deliberative and calibrated approach, and aggressively chasing for results. And finally, but most importantly, there is no substitute for business savvy.

In conclusion, the Railways's transformation is an exemplar for how state-owned enterprises can improve services despite all the challenges of balancing commercial and social objectives to fructify inclusive reforms. The following chapters will unpack the various attributes of the transformation strategy in substantial depth.

2 Political Economy of Reforms

INDIAN RAILWAYS, ITS STRUCTURE AND FUNCTIONING

Indian Railways is a unique state-owned enterprise because of its size, ownership structure, and 150-year history. These attributes, among others, make it a complex, intriguing, and thus fascinating subject. It is one of the world's largest infrastructure-providing state-owned enterprise. It is a ministry within the Government of India with 1.4 million employees and 1.1 million pensioners. It has one of the world's largest railway networks—over 63,300 km of routes—and runs approximately 13,000 trains each day, including 9000 passenger trains. It carries over 2 million tons of freight and some 17 million passengers between 7000 railway stations each day. This is achieved with a fleet of 200,000 wagons, 40,000 coaches, and 8000 locomotives. To fathom the scale, consider the fact that Indian trains, each day, travel four times the distance to the moon and back. The railway is vertically integrated and horizontally differentiated into functional silos. Under a single umbrella organization, Indian Railways finances, builds, owns, and manages most of its assets. This includes locomotives, wagons, coaches, rail tracks, stations, and enormous stretches of land, as well as hotels, schools, hospitals, and staff housing. Additionally, through a range of subsidiaries it manufactures and maintains most of these assets in-house (see Figure 2.1). This monolithic structure of the Railways has been a contentious issue among senior policymakers in the Government of India as well as international organizations.

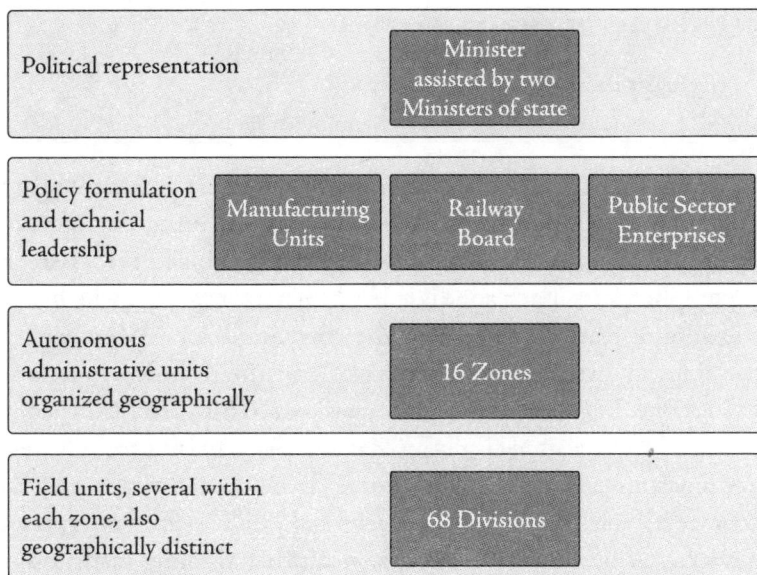

Political representation — Minister assisted by two Ministers of state

Policy formulation and technical leadership — Manufacturing Units | Railway Board | Public Sector Enterprises

Autonomous administrative units organized geographically — 16 Zones

Field units, several within each zone, also geographically distinct — 68 Divisions

FIGURE 2.1: Organogram of the Indian Railways

The apex body in the Railways is the office of the Minister, which brings with it a political mandate and associated leadership. This is followed by a three-level bureaucracy—the Railway Board, zones, and divisions (see Table 2.1). The bureaucrats are organized in a matrix of functional and geographic specialties. At the top is the Railway Board. It is composed of one member from each of the functional specializations of the Railways—electrical, engineering, finance, mechanical, staff, and traffic. The Board is led by the Chairman—better known as CRB or Chairman Railway Board. All members of this office rise through the ranks of the institution and thus have enormous experience and insight but short tenures—a year or two before they retire. Thirty-five directorates assist the Board in fulfilling its functions.

TABLE 2.1
Staff strength

Type of unit	Number of staff (2008)	Staff grade	Number of staff (2007)
Railway Board	1842	Group A and B	16,000
Manufacturing units & public sector enterprises	50,426 (44,426 + 6000)	Group C	907,000
16 zonal railways (including 68 divisions)	1,326,663	Group D	484,000
Total	1,378,931	Total	1,407,000

Zones are the apex bodies in the field and act as intermediate policy level. These are relatively autonomous units that govern the functioning of the Railways and are organized into sixteen geographic areas. Each zone is led by a General Manager who heads the administrative activities of the concerned zone. Members of the Railway Board provide oversight over the technical functions in the zone. Furthermore, each zone is parsed into several divisions, each led by a Divisional Railway Manager. The division is the lowest administrative level where the zonal departmental heads play an overseeing role. Finally, there are several public sector enterprises, manufacturing units, workshops, and other training institutions that report directly to the Board.

EVOLUTION

The inception of the railroad in India is often associated with the inauguration of the maiden journey of a fourteen-carriage train carrying 400 passengers at 20 miles per hour between Mumbai and Thane on 16 April 1853. But this was not the first railroad in India. In 1836–7, some seventeen years earlier, linked to a stone quarry, a 3½ mile-long railway track was laid in Chennai—then known as Madras Presidency. On this ran the first freight train in India. Yet another surprising fact is that it was powered by wind sails and consequently was called the *Wind Carriage Railway* as reported on 30 December 1837 in the *Madras Herald* (Bhandari 2006: 2). With looming concerns over climate change that beset the modern economic models based on consumption of fossil fuels, in hindsight the use of a renewable source of energy like wind was astonishingly progressive.

Much of the initial railroad construction was led by private firms including the East India Company. These constructions were financed through investments from capital markets in England backed by British government guarantees. The princely states of Bikaner, Gwalior, Jodhpur, among others financed their own railroads as well. Eventually, the crown realized that the incentive structure for the private contractors did not encourage parsimony. This was primarily because all risk was borne by the state that guaranteed a 5 per cent return on investment. Alternately, the contracts had provisions to buy back the infrastructure if it was unprofitable for the private firms that built it. Additionally, the state provided land gratis and required its mail to be carried free of charge. As a result the private contractors in some cases spent lavish amounts to construct railroads. Such public–private collaboration in the nineteenth century resembles the contemporary cost-plus models with an assured rate of return. To counter the private disincentives, and respond to security concerns due to the revolt of 1857, the British government decided to take on the task of railway construction and management. Fast forwarding to Indian Independence in 1947, there were about forty-two railways that were all nationalized and consolidated into one state-owned enterprise. The timeline in Figure 2.2 illustrates the pivotal events that summarize the evolution of the rail industry in India.

Indian Railways has an affinity for technological development and possesses the in-house engineering prowess to keep pace with progress in the global rail industry. At the risk of oversimplification, let us assume a railway to consist of only four important components that constitute its capital stock. The all familiar railway tracks, the carriages that roll on them, the engine that pulls these carriages, and the signalling systems that tell the engine driver when to start and stop. After 1947, during the post-Independence nationalization and resultant consolidation of Indian Railways, several types of

Technological Evolution

1836	1891	1925	1967	1971	1992	2002
A short railroad built near Chintadripet, Chennai. Train powered by wind.	3rd class coaches get toilets.	First electric train service, Mumbai to Kurla.	Cement concrete sleepers introduced.	Policy adopted for gauge conversion. Two years later steam engine production terminated.	Adoption of unigauge policy.	Jan-Shatabdi train launched.
1853		**1936**	**1969**	**1984**	**1998**	**2006**
Bombay to Thane passenger train service inaugurated on 16 April.		Air-conditioned coaches arrive.	Rajdhani Express makes maiden journey, Delhi to Howrah.	Introduction of first Metro Rail system in Kolkata. Next year, computer-based passenger reservation introduced. Shatabdi begins service 1988.	Konkan Railway begins operations.	Garib Rath and e-ticketing initiated.
			1950			
			First indigenous steam engine made in Chittaranjan.			

Institutional Evolution

1849	1905	1947	1974	1998
Financial guarantee for private railways that construct and operate—5 per cent return on investment, buyback policy, social obligation of transporting official mail and military, land given gratis.	Railway Board formed.	Indian Independence and nationalization of 42 railroad companies.	Rail India Technical and Economic Services (RITES), a consultancy unit created. Two years later. Indian Railway Construction Corporation (IRCON) created.	Guinness Certificate for Fairy Queen, world's oldest working steam Engine 1855.
1887	**1924**	**1950**	**1979**	**1999**
Victoria Terminus built.	Ackworth Committee recommends separate budget for railways.	Railway organized into 6 zones, Central Advisory Committee endorses.	Central Organization for Railway Electrification (CORE) created.	World Heritage site status for Darjeeling rail. Guinness Certificate for largest Route Relay Interlocking System, Delhi. And Indian Railways Catering and Tourism Corporation (IRCTC) created.
1890	**1930**	**1957**	**1988**	**2002**
Indian Railways Act passed.	Central Standards Office, technical standard enforcing agency established.	Research, Design, and Standards Organization established to consolidate all technical standard enforcement.	Container Corporation of India (CONCOR) created. Next year, Indian Railway Welfare Organization (IRWO), formalized.	7 additional railway zones created, making the total 16.

FIGURE 2.2: Evolution of the Indian Railways

Source: Bhandari (2006).

rail technology were inherited from the numerous independent regional railway enterprises. Enormous effort was put into standardizing this uneven capital stock.

Operational Crisis of the 1980s

In the period following nationalization of the Railways, there were some critical technological improvements. The old stock consisted of less efficient and high maintenance technology. The brake system was vacuum based, and most engines were steam powered. Wagons had unreliable plain bearings and four wheels—implying shorter wagon length as well as lower load-carrying capacity. Further, screw coupling, the device that linked wagons to form a train, was manually operated and had limited strength, so forming longer and heavier trains was difficult.

The new stock consisted of better technology. The new diesel and electric engines were more effective because they could pull heavier loads, used less energy, provided greater operational flexibility, and were less polluting.[1] Wagons had lower-friction roller bearings and eight wheels—namely, covered and open BOX and BCX types—with greater volume and load-carrying capacity. The centre buffer coupling[2] device was also better. In essence, through a combination of such technological improvements, the Railways had better rolling stock. However, as the lifespan of the rolling stock is between thirty-five and forty years, old stock continued to operate along with the new.

Further, goods were accepted in both wagon and train loads. This created operational inefficiencies. Piecemeal freight required frequent en route marshalling, examination, and formation of trains. At an interval of every 400 km the Railways maintained goods yards for this purpose. This was complemented with a logistical network to repack small parcels en route at repacking sheds before they are finally sent to their destinations. To further complicate matters the old and new stock of wagons was mixed and forming a train required linking incompatible types of couplings via a bridging device—known as a 'baby coupling'—which was perpetually in short supply as it was frequently stolen and resold. The repeated reconfiguration of rakes—a set of wagons that form a train—required reissuance of new brake power certificates that needed frequent inspections leading to delays. Additionally, the steam engine required repeated halts for operational needs like coal and water refills and change of crew.

The old and new stocks of wagons with different types of couplings and bearings were jumbled to form freight trains. Frequent shunting was required because of piecemeal movement of wagons. This was a critical obstacle in the functioning of the railway system. As a result, the entire railway system was reduced to operating on the strength of the weakest link. Railway sidings—these are auxiliary tracks—were cluttered with 'sick' wagons and yards had become bottlenecks, primarily due to wagons waiting to be remarshalled and transferred to other trains. Furthermore, there was a union of engine drivers which was making tough bargains. In sum, these and other operational and management practices caused uncertainties and long delays in the entire freight operations—power plants awaited coal supplies and transportation of essential commodities such as foodgrains and petroleum products were severely constrained.

On 17 November 1980, in a dramatic change of fate, the entire Railway Board was replaced at the behest of then Prime Minister, Indira Gandhi. She identified M.S. Gujral who became the first General Manager in the Railways to be directly appointed as Chairman of the Board. Gujral took bold steps on arrival. He had a twofold strategy. First, he segregated the old and new types of wagons—specifically, four wheelers from eight wheelers, screw coupling from centre buffer coupling, and roller bearing from plain bearing.

The steam engines were utilized on short routes, like operating trains between yards and shunting. And for the long haul higher horsepower engines were deployed. As a corollary to the separation of stock, Gujral improved maintenance practices at the start of a train so as to abolish the practice of en route examination of trains called 'safe to run' at every 400 km (and 'intensive' at every 800 km for intensive routes). This was replaced by end-to-end examination. Second, to further improve operational efficiency, he terminated the practice of accepting less than train load freight and introduced the concept of 'block and point-to-point trains', thus eliminating the need for trains to halt en route. He revised the unit of transportation from wagon load to train load. Earlier, for every change of steam engine new brake power certification was required through a safe to run examination. Further, train examinations were required in case of re-marshalling of trains because the unit of transport was a wagon load. However, now trains no longer required renewal of brake power certification for change of locomotive at short distances. He further ordered the elimination of several redundant yards. As a result of these two actions, the time spent on shunting wagons at yards or the need to halt at every yard, was reduced.

As long-term measures, Gujral introduced improved brake, bearing, and coupler technology in all wagons as standard practice. Second, production of higher-capacity and better-designed air brake BOXN and BCN wagons was initiated, high-power diesel locomotive production was ramped up, and high-capacity diesel-powered breakdown cranes were acquired. Third, he prioritized the electrification of railway routes, improved utilization of diesel and electric locomotives, and ordered the complete phasing out of steam engines.

In sum, the Gujral reforms resulted in a quantum jump in the Railways's operational performance. Not only did trains roll faster but they also increased the amount of freight transported. There was a fourfold increase in the incremental freight carried in the decade following the reforms compared to the preceding decade. Gujral had also explored the need to increase axle load—essentially carrying greater amounts of load in a freight wagon. On his rather abrupt departure, risks associated with the decision took precedence, resulting in no further follow-up for the following two decades. The present railway transformation leveraged the new rolling stock that Gujral had introduced through his long-term plans.

FINANCIAL CRISIS IN 2001

In 2001, Indian Railways faced a severe financial crisis. It defaulted on dividend payments to the Government of India, its cash balance shrank to a paltry Rs 359 crore, and the Railways did not earn enough to be able to replace aging assets resulting in large replacement arrears. The profitable freight business was recording a poor growth rate of 3 per cent and its expenses grew faster than revenues. The Railways's financial condition was unsustainable and it was on the verge of bankruptcy. There were a range of internal and external factors that led to this deteriorating operational and financial condition of the Railways. In response to the currency crisis in 1991, the Government of India initiated liberalization of the command and control economy; this marked the retreat of licence raj. These reforms reversed Fabian economics by reducing barriers to trade—revoking quotas, licences, permits, as well as reducing tariffs on imports of intermediate and finished goods. Additionally, with deregulation of internal markets, restrictions on large and small firms were gradually repealed. As a cumulative effect of these liberal reforms, firms began feeling competitive pressures from domestic and international firms because in a liberal trade regime domestic prices of tradable goods and services converge with global ones. This is particularly true since the cost of international freight transportation has been declining. In response to stiff competition from domestic and international firms, producers began reviewing their cost structures, including total logistics costs—namely cost of transportation, inventories, multiple modal transfers, delays, and damages.

In the pre-reform era, under the freight equalization scheme, the cost of transportation of crucial bulk commodities like steel and fertilizers was neutral to the lead of transportation as the difference was paid by the public exchequer through a subsidy. The freight equalization policy for steel was the reason that steel made in Jamshedpur cost about the same in Ranchi as in Gujarat in those days. Further, oil pool account for petroleum products and the retention pricing scheme for fertilizers played a similar role. In essence, the producers of these commodities were not concerned about the costs associated with transportation because they could pass on these costs to the state. Liberalization began dismantling this arrangement. Hence firms became cost conscious and began seeking cheaper transport services. Since transportation of bulk commodities lay at the heart of the Railways's post-Gujral phase, liberalization was calling into question this freight business model as customers migrated to alternate modes.

Moreover, a fiercely competitive private road transportation sector was increasingly acquiring railway's market share. There were other competitors on the horizon as well: international logistics firms, shipping industry, and oil pipelines. Further, the Railways experienced another external shock from the reformed macroeconomic environment. There was a sharp decline in the ability and willingness of the central government to provide budget support through fiscal transfers for capital investment needs or recurring expenses like the increase in wages due to the fifth pay commission. The former enhanced competition from the extremely demand-responsive private road transport market and the latter eroded hopes of bailouts through fiscal transfers.

Finally, over the 1990s, US$ 7 billion (Rs 30,000 crore) was invested in improving the quality of tracks.[3] The new tracks were stronger and could endure heavier loads. Additionally, by the turn of the century, billions of dollars were invested in acquiring racehorse-like engines—relatively expensive high horse power diesel and electrical locomotives. Yet the modernization efforts and inherent strength of the bureaucracy were not translated into benefits despite huge demand for freight services because of a lack of synergy among railway departments and a missing commercial focus, among other things. These constraints will be discussed in subsequent chapters. In sum, while there were significant technological and institutional improvements over two decades, it did not lead to tangible improvements in productivity. As Lalu describes the condition, they had acquired a Jersey cow, but did not milk it adequately, resulting in a sick cow.

Beyond Bankruptcy

The financial condition of the Railways was so precarious that the Government of India convened some of the brightest policy-makers and private sector experts to diagnose and advise on corrective measures. Rakesh Mohan, noted economist and presently Deputy Governor of the Reserve Bank of India was Chairman of this expert group. The combined intellect of the Government of India's expert group on railway reforms as well as global experts pegged this near bankruptcy scenario on the 'split personality' of Indian Railways. The conflicts in achieving multiple organizational goals—welfare and commercial—were to be blamed. In the expert view, there were five essential contradictions that the Railways needed to resolve. First, there was the political mandate that led to conflicting priorities between the politician and the bureaucracy—the Minister and the Railway Board, railway's top management. The experts argued that Indian Railways was heading towards bankruptcy because ministers meddle with financial allocations resulting in poor choices of investment in politically motivated unremunerative projects. Further, the planned economy notions that saddle the railroads with social obligations like cross-subsidization of passenger fares through frequent increase in freight tariffs that is eroding the market share of this profitable segment of business. This cross-subsidy is reflected in the fare to freight ratio[4] in Indian Railways, which is one of the lowest in the world. Furthermore, there was concern that even within the passenger segment the potentially lucrative premium class passenger segments like the air-conditioned coach travellers were taxed—because they paid higher prices—in order to subsidize the ordinary sleeper classes. As these premium class fares were on the rise, the Railways was losing these customers to budget airlines.

Second, since the policymaking, oversight functions, railway ownership, and management are all concentrated in a monolithic organization, there is lack of accountability.

Third, the Railways indulged in a variety of non-core activities that ranged from in-house manufacturing and maintenance of engines, and carriages to even catering—remember Rail-neer?[5] Additionally, it was burdened with the social obligations of running hospitals and schools, *yatri niwas* hotels, training institutions, and providing employee housing. These distracted railway men and women from focusing on the core business of running trains. These non-core businesses belonged to a bygone era and were not in keeping with contemporary practices.

Fourth, there was a fiscal crunch due to declining budgetary support[6] through central government transfers. The budgetary support has declined threefolds from 75 per cent in the fifth plan (1975–80) to 25 per cent in the ninth plan (1997–2002). To make up the gap in investment needs, the Railways borrowed from the markets.

Finally, there was the army of rail employees and their ever-increasing salaries and pension liabilities. Where staff costs accounted for about half of total costs, the implementation of the fifth pay commission recommendations would act as the last nail in the fiscal coffin.

In sum, a combination of political interference, conflicting commercial and social objectives, fiscal crunch, lack of market incentives, and unproductive employees had hindered investment in track renewals and other safety measures, while the Railways was losing market share.

In response, the Mohan Committee and international experts recommended institutional restructuring, stating that 'at present, IR faces two possibilities: significant change through reform, or a financial and operational collapse' (Sondhi 2002: 37). Based on the preceding analysis, the experts carefully crafted a reform package: (1) unbundle the institution into separate roles—policy-making, regulation, management—by corporatizing railways and establishing an independent regulator, specially for tariff setting; (2) privatize non-core activities like health care, education, production, and maintenance of trains; (3) reduce the 1.6 million staff by 25 per cent; (4) reduce cross-subsidy, hike fares for second class passengers by 8–10 per cent every year for five years; and (5) separate social and commercial obligations.

These reform recommendations received an emotionally and intellectually charged response. Labour unions held a *dharna* and raised the red flag in Kolkata (*Frontline* 2001). There was unease about the organizational restructuring among top management of the Railways. As a reaction to the Mohan Committee reforms, the Railways's management responded with a status paper that recognized the need for change. The status paper tabled the following five central measures of action for consideration of parliament. It proposed to corporatize non-core activities citing the past success with its consultancy, container, construction, catering, and telecommunication business subsidiaries. It aimed to set clear targets for reducing staff strength to the 'right size' from 1.545 million to 1.18 million by 2010 through natural attrition. Rationalization of freight and passenger fares was considered. Regarding cross-subsidies a case for interest free transfers from the central government was made. Finally, to address the fiscal gap it was proposed to consider loans from multilateral banks—World Bank and Asian Development Bank— as well as co-financing with state and local governments and cautious leveraging of private equity.

The status paper was mute on core restructuring of the vertically integrated monopoly structure of the Railways. Instead it favoured the devolution of more discretionary power to zonal level for approval of capital investments projects. Thus, general managers could make larger decisions about the priorities in their zones, but were not unbundled into autonomous competing units, as was envisaged by the Mohan Committee Report. Many of these efforts—tariff rationalization, reduction in staff strength (by not filling two-thirds of job vacancies), borrowing from the World Bank and Asian Development Bank—were implemented. While these efforts brought some respite from the fiscal crisis to a beleaguered railway bureaucracy, the condition of the Railways remained precarious.

Bit of a Contradiction

The international expert (Thompson 2003 and Sondhi 2002) and Mohan Committee (2001) recommendations[7] were conventional wisdom but many of them presented the antithesis of the populism that Lalu Prasad stands for. Arriving at Rail Bhavan, headquarters of the Ministry of Railways, the minister did not disappoint his supporters, or critics. His position on several policy issues was the converse of the recommendations proposed by the Mohan Committee and other international expert groups. This contrast is captured in the following significant steps taken by the Railways since he became Minister of Railways.

While the experts had recommended retrenchment, the Minister planned to use the Railways as a vehicle to generate employment, both within and outside the ministry. As a first step, he banned the use of plastic cups on railway stations and trains, replacing them with *kulhad*s or clay pots. Similarly, synthetic upholstery and linen for offices, trains, stations, and *yatri niwas* were to be replaced with khadi handspun yarn, and hand-woven cotton cloth. The purpose was to increase employment for the rural artisans and khadi handloom weavers through the use of handmade cups and cloth. Finally, he hired 20,000 coolies as railway staff for the post of gangmen—class four employees and frontline workers for railway track maintenance.

The experts had recommended a fare hike in the loss-making passenger segments and establishment of a tariff regulator. In contrast, the Railways did the converse and reduced fares in each budget and in every travel class—from air-conditioned coaches to unreserved passenger coaches—and by at least Rs 3 (7 cents) for poor passengers.

Further, the minister was of the opinion that as an elected representative of the people, if he was in-charge of the Ministry of Railways and answerable to the people via Parliament, he could not devolve the role of determining tariffs to an auto-nomous entity—an independent tariff regulator. Thus, he wanted independence from a regulator in deciding tariffs as opposed to independent regulation.

The Mohan Committee and international experts had recommended corporatization and divesture from non-core businesses. But the minister planned to build new production units to manufacture diesel and electric engines, wheels, and passenger coaches. Three factories were to be built in his constituencies: the rail wheel factory in Chhapra, a diesel locomotive factory in Maruhara, and an electric locomotive factory in Madhepura, all in Bihar. The fourth, a factory to manufacture rail coaches was to be built in Rae Bareli, the constituency of Sonia Gandhi, the chairperson of the United Progressive Alliance (UPA).

Moreover, the Railways acquired sick freight wagon factories like the Mokama and Muzaffarpur units of the Bharat Wagon and Engineering Company in Bihar from the Ministry of Heavy Industries, Government of India. Additionally, it acquired the scrap and land of the Dalmianagar industrial complex in Rohtas, Bihar, to build factories for manufacturing essential components of wagons. While the expert group deemed such investments a distraction from the core business of transportation (Mohan 2001: 9), Lalu saw them as a political necessity.

The Mohan Committee had criticized investments in unremunerative projects like the construction of new railway lines, urban rail transportation, and the uni-gauge policy.[8] However, the Railways announced the conversion of the entire 13,000 km route from metre gauge to broad gauge to be completed by 2012. Between 2005 and 2008, twice the number of new railway lines have been sanctioned—41 new projects approved at an expense of Rs 10,500 crore (US$ 2.4 billion). Additionally, the experts expressed concerns about the Railways giving priority to loss-making passenger trains, as opposed to securing track space for profit-making freight trains. Yet, 1500 additional passenger train services were announced between 2005 and 2008, a 10 per cent increase over the previous four years. Finally, while the experts had recommended that the Railways should isolate its total social burden and seek central government subsidies for the same, the Railways substantially added social obligations, even though the government lacked the fiscal space and willingness to offer subsidies.

In conclusion, the expert group recommendations were text-book solutions for restructuring the Indian Railways—unbundling and separation of social and commercial functions, retrenchment, independent regulation, corporatization, and fare hikes. Much of the costs of proposed reforms would pose a burden on the common people and railway staff, at least in the short term. But this did not resonate with Lalu, a populist politician concerned about the masses—remember the 300 million people in India that live on less than Rs 18 a day?[9] Yet the minister wanted the impossible. He made a bold announcement, aspiring to make Indian Railways the 'world's best'.

Not surprisingly, the Minister's announcement was met with contempt. Public opinion expressed in the media, and the in railway's bureaucratic inner circles, was rife with scepticism about the Minister's credentials as well as his intentions. Lalu was ridiculed because of the negative perception of his past performance in Bihar where he and his wife had led the state for a decade and a half. In the first few months of the Minister's arrival, these apprehensions seemed to crystallize in two striking incidents at Rail Bhavan, the Railways headquarters in Delhi.

Soon after his arrival at Rail Bhavan, Lalu was visited by a Member of Parliament (MP)[10] from the state of Uttar Pradesh. During the meeting the Minister called the Chairman of the Railway Board, the seniormost bureaucrat in the Railways, and asked him to consider the requests of this visitor. Soon after the meeting was over the Minister left for Parliament. The visitor decided to pay a visit to the chairman whose office is next door. The visitor and the chairman had an unpleasant exchange. The visitor accused the chairman of being a political puppet of the previous government and ridiculed him in unmentionable words. Understandably, the chairman was furious, and perhaps felt insulted and humiliated. He wanted to proceed on leave immediately. In the railway bureaucracy, leave of this nature implies intent to resign.

Next, another MP requested a car from the ministry. Even though he was not entitled to one, the Ministry of Railways sent an air-conditioned white Ambassador.[11] However, this influential politician was offended by the car he received. He expressed his displeasure and stated that the use of such an old-fashioned car was not commensurate with his stature. He requested the vehicle to be replaced with a more elegant and comfortable Honda City. The railways administration in-charge of cars complied.

In sum, such events seemed to confirm the public fears of Rail Bhavan morphing into 'Bihar Bhavan' and jungle raj displacing *rail raj*. It was a turbulent time in the Railways headquarters, with rumours rife in the corridors of power. The media followed suit, newspapers and news channels were generous in their daily pessimistic reporting of the Minister, his family, or his political party.

Earning Trust

Little did the railway staff or media realize that in his new assignment Lalu was going to dispel the image associated with his leadership in Bihar. The Minister was determined to demonstrate his administrative acumen and rustic common sense. In a test of leadership, the Minister stood firm.

When the Minister learned of the developments with the chairman, he left the Parliament session and rushed back to Rail Bhavan. In the dramatic meeting that followed, Lalu clarified his personal stance, declaring zero tolerance for political interference with railways's daily operations. He encouraged the chairman to take decisions based solely on railways rules and regulations and further assured the chairman that he had complete confidence in him and promised wholehearted support. In this moment, the chairman was won over as the first and the most powerful ally who would champion the future rail reforms.

Regarding the second incident, the Minister is briefed about the car request and to everyone's surprise he is upset. Not only does he inquire why such a request was entertained, but he also takes institutional action by issuing a memorandum. In the memo, senior railway managers are instructed not to dole out favours to individuals who claim to seek these under the Minister's patronage. Additionally, any notes or telecommunication claiming to be ministerial instructions must be clarified with the Railway Board and instructions sought in writing. These instructions were reiterated every few months for a period of two years because habits die hard. Thus a zero tolerance towards breaking of rules was enforced. In the interim, many misdemeanours were registered against the Minister's family members, party workers, and other political allies. These violations ranged from travelling in a higher-class coach than the ticket warranted, to requests for changing the platform on which a train was scheduled to arrive. Despite stiff resistance from influential quarters, the Minister stood firm and consistently backed railway staff, allowing them the room to implement rules and regulations, charge fines, and take corrective action.

By setting aside personal agendas and keeping his word, the Minister earned the respect and trust of colleagues. For Lalu, this notion of self-discipline was extended to include his family, party workers, and political allies. The value of such self-discipline by the leadership is best captured in Peter Drucker's observations, 'I no longer teach the *management of people at work*', instead 'I am teaching, above all, *how to manage oneself*' (Pandya and Shell 2005: 63, emphasis added).

Further, the Minister demonstrated a non-partisan ethic in his daily engagement with the Railway Board. There are two types of responses from the Minister that are worth noting. First, senior bureaucrats in the Railways were concerned that the Minister would favour candidates based on their social sub-groups (castes) as opposed to merit. But to their surprise, he adopted a hands-off approach allowing the management to select their own teams based on meritocracy.

Second was the transparency in approval of competitive tender bids. The Minister upheld and promptly approved the recommendations of the Board on the award of tenders in construction works and procurement contracts that ranged from a few hundred to thousands of crore rupees.

In this context, understanding of the distinct and complementary roles of political leadership and the permanent civil service is essential to the functioning of a ministry. These roles are well articulated in the Indian Constitution.[12] The politician comes with a mandate from the people—such as no privatization, no retrenchment, and no fare hike—which sets the broad policy agenda. The bureaucracy, within the limits of the law, is the implementation agency. However, in practice these roles tend to clash. Some politicians interfere with implementation, especially in transfer, posting, new staff appointments, issuance of contracts, and other operational matters where preferential treatment results in political gains. And bureaucrats deploy their discretionary powers to stall the political mandate by avoiding or slowing down execution. The mutual contempt between the political leadership and the civil service reinforces unproductive behaviour. Conversely, a politico-bureaucratic arrangement is productive when the politician does not meddle in the functioning of the executive and in turn the bureaucracy accepts the political mandate. In the railway reform, a prerequisite for this arrangement was making a distinction between the 'political mandate' and petty 'political interference' in routine administrative activities and then developing complementary roles.

Through routine engagement between the bureaucracy and the politician a sense of trust, respect for differences, and understanding emerged. There were mutual differences, and these persist, but the bureaucracy learned to respect the political mandate and in return the political leadership maintained a hands-off approach to railway's operations. This may appear presumptuous, but it is difficult to rationalize how else an organization close to bankruptcy in 2001 would reap a cash surplus of Rs 25,000 crore (US$ 6 billion) in 2008. While integrity of the leadership is the bedrock for the transformation of the Railways, it alone is not sufficient. To formulate and execute the reform strategy it required a core team of reformers with a deep sense of purpose, a sharp understanding of the political economy, and business acumen.

Analysing the Political Economy

Lalu likes to challenge his staff with his rustic references. In the early days, he often asked the management to explain why, if a cowherd like him could produce a profitable herd of five hundred from a few cows, the Railways with its 200,000 wagons, 40,000 coaches, and 8000 engines was running at a loss? But he neither sells the cows' milk at a loss nor does he overstaff the barns. The veterinary services for the cows, education, and health care for his staff are not provided in-house. Besides, the milk produced is sold at competitive prices to a premier five star hotel in Patna. The Minister acknowledges that the Railways's challenge was a greater one because, unlike him, the Railways has several social obligations.

Before proceeding with an analysis of these conflicting objectives and their implications for reform, let us briefly review what makes a profitable commercial organization. Consider a simple example of a firm that sells goods or services. A firm is profitable if its costs are less than the selling price because profit is the difference between selling price and costs (profit = price − cost).[13] But if costs are greater than the price then the firm makes a loss.

To transform a loss-making business into a profitable one, private firms respond in one of three ways. Jack Welch, Chairman and CEO of General Electric (1981–2001), aptly summarized this strategy as the 'fix, sell, or close' options (Welch and Byrne 2001: 111). To 'fix' a loss-making business, either the price of goods sold needs to be increased or the costs decreased.[14] But if the business cannot be fixed through these measures, then one should 'sell' those aspects of the business that are loss making. Third, if fixing or selling is infeasible, the next step is to 'close' a business and sell the residual as scrap. Welch had devised this strategy in response to two questions posed to him by management guru Peter Drucker, 'If you weren't already in a business, would you enter it today? If the answer is no, what are you going to do about it?' Welch responded with the target of making each business in General Electric the number one or two in the world, else they needed to be fixed, sold, or closed (*Business Week* 2005).

In the private sector, this strategy works well—as it did for General Electric, IBM, and a host of others. But in the thick socio-political reality of the Railways such a strategy is at best economically desirable but politically infeasible. This infeasibility is deep-rooted because expense cutting is accompanied by the high social costs associated with restructuring—retrenchment and disinvestment. Thus the challenge confronting the policymakers was to deliver inclusive reforms—cater to all the unremunerative social obligations and yet transform the Railways into a commercially viable organization of global repute. The challenge was daunting, but the Minister and his advisors along with the railway management took on the task. Transforming the Railways through inclusive reforms became the mission around which the Railways's core strengths were deployed (Budget Speech 2006: 1). This called for a creative response. Past assumptions and practices were to be reviewed. Therefore, the core team of reformers began with analysing the reform context.

CRAFTING THE SPACE FOR REFORM

The core team discovered that to understand the political economy and identify space for reforms within the Railways, all policy initiatives must be screened at two levels. First is a market test, where the consideration is commercial viability. Second is a societal value test, where the evaluation is based on the political desirability and feasibility of the initiative. Reforms that pass the first test but fail the second are commercially desirable but politically infeasible. But the reformers accepted the social considerations as integral to the political mandate and with the curiosity of policy geeks plumbed the depths of the challenge. They considered a step-by-step approach. Each variable that affects the Railways's functioning was split into its apolitical and political components, across scales and dimensions—from the macro to the micro. The core team was determined to look beyond the obvious. Their approach to traversing the complex political economy of the Railways involved a lot of learning by doing. An incremental approach that can best be described as muddling through. But the five step analysis presented here captures the essence of the strategy. The primary objective of this experiment is to assess the true extent of the Railways's 'split personality'—conflicting commercial and social goals, and to seek to verify if these conflicts were the principal obstacles to the deteriorating condition of the Railways. In the process, the space for reform was crafted. The results are startling!

A macro analysis to separate the political aspects from apolitical ones is the first step. In each facet of the Railways's functioning— from pricing, commercial operations, and investment policy to labour management—every policy initiative was subjected to the twin test along commercial and social lines. This has to do with discovering the proportion of investments and revenue that are considered politically sensitive due to their social implications for the poor at macro level, as against those with exclusively commercial implications.

As seen in Chapter 1, only 20 per cent of the total capital investments of the Railways have political implications—the socially determined investments. The rest 80 per cent investments have little political repercussions. Instantaneously, an enormous apolitical space for leveraging investments was created. These allocations can be market-oriented and based on operational and commercial requirements such as route-wise planning of congested high-density networks where demand far exceeds supply. But route-wise planning requires all departments to channel their efforts into getting all the technical and operations details right—from tracks, stations, and signalling to scheduling and running trains. Despite discussions on this matter for decades, the Railways was unable to act, not because of political interference, but because of deep rifts between functional silos that are organized into departments.

Further, total traffic earnings were disaggregated in a similar manner. A quick glance suggests that 22 per cent of the revenue is from market segments that are politically sensitive—largely second class passenger trips in reserved and unreserved coaches that are patronized by poor people—an outcome again with exclusively social return. The rest 78 per cent of the revenue is earned from the apolitical segment of customers—all freight, parcel, high-end passenger segments and other sundry earnings. This business segment can—and should—be managed on commercial principles, but the Railways lacked a business orientation in the past. This was yet another startling finding. As these variables are analysed further the space for reform continues to grow.

The social costs of retrenchment, divesting, and other cost-cutting measures are high and so they are politically infeasible. Therefore, the Railways management lacks control over many operating expenses. However, there is no political constraint in distributing the same operating expenses over larger volumes to reduce unit costs. For example, the number of employees cannot be reduced, but if trains carry more goods or there are more trains with the same number of employees, unit costs decline. Similarly, while at macro level passenger fares are politically sensitive, the yield per train is apolitical. Thus, increasing revenues through passenger fare hikes and reducing unit costs through retrenchment were thought to be outcomes with exclusively economic returns but this analysis demonstrates how they have been converted to win–win outcomes.

As a second step, the effort was to identify apolitical sub-components within the politically sensitive business segments. This is because the reformers had not given up on the potential of the remaining 20 per cent in investments and 22 per cent in revenues that were identified as political at macro level. This 20 per cent of investments comprises of 8.75 per cent for new railway lines, 1.59 per cent for urban transport, and 9.05 per cent for gauge conversions for the fiscal year 2008. Let us consider gauge conversion. The experts branded this as an unrenumerative investment and perhaps they were right. Yet, in a piecemeal manner, more than half of the gauge conversion is done and can be leveraged for remunerative activities. While different gauge conversion projects were at varying stages of completion, the reformers focused on projects that enhanced the capacity for the lucrative freight business. Alternate routes for congested railway lines were discovered through route-wise planning of these gauge conversion works. Nearly half of the gauge conversion projects were helpful in capturing additional freight business and were therefore commercially viable. Some examples of such alternate routes are links to ports and connectivity to cement plants, marble, and granite quarries. Essentially, through such an analysis the initial 80 per cent of apolitical investments expanded to 85 per cent because some components of the gauge conversion were included.

Similarly, let us consider the 22 per cent of revenues from the passenger segment that had been set aside as political. While passenger fares for suburban rail service or second class passenger trains are politically sensitive, non-passenger revenue is apolitical. Consider the Victoria Terminus and the slew of stations along the suburban railway network in Mumbai. With 7 million dedicated customers each day, the railway stations have access to the gaze, the footfalls, and refreshment needs of these hard-working Mumbaikars. Quite like the fabulous advertisement and sponsorship revenues that account for crores of rupees earned in the Indian Premier League cricket games, railway stations can leverage their strategic position to earn without raising passenger fares. Such strategies allow the Mumbai suburban service of Western Railway zone to remain profitable despite passengers being charged below cost fares. As the reformers segregated these apolitical aspects from political variables, the space for reform further expanded.

Third, this space for reforms further widens when perceptions were reconciled with reality. The railways had a uniform policy to subsidize all customer segments, based on the principle of a customer's perceived affordability—low-value commodities were charged cheaper rates. Emphasis is laid on perception of affordability because in practice there was no explanation for why customers who transported low-value bulk commodities like iron and manganese ores were charged less than cement and steel customers. Surprisingly, this was the practice at a time when commodity prices were booming and there was no political compulsion for keeping prices low. Additionally, there was a firm belief among the Railways bureaucracy that the transportation of foodgrains and fertilizers needed to be underpriced. They assumed an increase in freight would be politically undesirable because it would result in price increase for the poor consumers like small farmers. In reality, the price for foodgrains and fertilizers is fixed by the Government of India and the differential between the cost and fair price is borne by the government. When the Minister was briefed that increases in freight charges would not result in increased food or fertilizer prices for the consumer, he readily agreed to revise freight charges for these commodities. This is an example of how what was thought to be an exclusively social outcome turned out to be a win–win outcome—win–win because the Railways increased its revenue while poor consumers were not affected. In sum, such gaps between perception and reality were demolished and were crucial in further crafting the space for reform.

Fourth, each policy initiative—pricing, commercial, and operational—was broken-down into its nano constituents and reconfigured into the four outcomes. Table 2.2 illustrates how political variables for each policy issue in every business area are separated from apolitical ones.

In conclusion, the spirit behind the analysis was to accept the political engagement not as interference but as the democratic mandate and work towards an inclusive transformation. In return for the bureaucracy's acceptance of the political mandate, the political leadership had to earn trust, as it did with self-discipline. With this as a base for reforms, a relook at the hidden opportunities behind the thick veil of the political economy led to some striking discoveries for the reformers.

Within the existing structure of this bureaucracy, the analysis revealed that the space for reform was extensive. Moreover, the analysis targeted the elimination—or at least the contraction—of the gap between perception and reality. This was done policy by policy, business by business, and activity by activity. Several outcomes that were perceived to be exclusively economic or social in their objectives were found to be both. Thus, in practice, they belonged to the much

TABLE 2.2
Disaggregating variables into political and apolitical

Policy	Political	Apolitical
Pricing	Passenger fares for non-air-conditioned classes—reserved and unreserved second class. Freight charges for salt.	Passenger fares for air-conditioned classes. All commodities other than salt which is 99 per cent of freight business.
	Parcel charges for fruits—banana, orange, mangoes.	All commodities other than fruits which are 90 per cent of parcel business
Commercial	Affordability and honour-based concessional fares—for senior citizens, students, farmers, freedom fighters, army, and the like.	Over 90 per cent of non-passenger income from catering, parcel, land lease, and advertisement revenue.
Operational	Priority to less remunerative trains. For example, profitable freight trains give way to passenger trains.	All other operational decisions.

Source: Authors's compilation.

desired win—win outcome. The apolitical space for reform was further enhanced by this micro disaggregation of various business segments, leaving as much as 85 per cent in investments and 80 per cent in revenue for commercial considerations. Clearly, in the analysis the primary constraint for the Railways's productivity was not political interference but rampant departmentalism (Tandon 1994), poor commercial orientation, an aversion to profit, lack of customer focus, and a monopoly mindset despite shrinking market shares (Mohan 2001a).

The problem was then less psychic and more physical. As observed by the Tandon (1994) expert group, the organizational structure of functional silos routinely compromised institutional goals for departmental gains.

Departmentalism, its all-pervading culture, is regarded a 'bane and a curse' of the system in which each discipline or area becomes an 'Island unto itself, no part of the main'. Consequently, a hierarchicalised caste system prevails in which some departments are considered superior to others. It weakens the organization, notably because in the absence of a team spirit, owing to departmental loyalties, the interests of the organization are subordinated to a department's own narrow considerations. In consequence the organization suffers while the 'whole becomes less than the sum of the parts'. This is particularly so when projects are perceived and pushed in the interest of a department even at the cost of the organization [p. 8].

Therefore, the focus of the reforms shifted from no privatization, no retrenchment, and no fare hike to identifying, expanding, and multiplying mutually beneficial outcomes and leveraging them to maximize financial returns without adding social costs. How this was achieved is the subject of the following chapters that unpack the market response into demand- and supply-side strategies as well as a public sector management approach. As an immediate step, care was taken to minimize unproductive outcomes by inducing synergies between departments, establishing cross-functional teams, encouraging project-focused coordination with well-defined targets under the close eye of senior management.

Strategy

Having identified the space for reforms, the core team needed to act by setting ground rules for engaging in the political economy. In politics, perception often counts for more, reform pace (or the lack of it) matters, and form precedes substance. Thus the strategy had three elements—political stance and phraseology, incrementalism, and tact.

In order to inspire confidence among interest groups—media, parliamentarians, bureaucracy, labour unions, among others—the Minister had announced, 'No privatization, no retrenchment, and no fare hike'. It was a bold statement. Here the choice of words was important. Such a statist stand continues to appeal across party lines, and received thumping applause in the Parliament. The bureaucracy felt reassured with the no sell-out, fix, or fire stance. As a next step, the Minister qualified his statement. He was against privatization of core railway functions, but was not against (meaning reluctantly in favour of) public–private partnerships in non-core functions of the Railways. The Planning Commission too had been recommending the introduction of private operators for running container trains for several years. On the other hand, the left front also did not oppose this stance because of Lalu's broader political image of being a staunch supporter of the public sector and the poor. Furthermore, the Minister qualified his stance on private participation. The core operations of track maintenance, hauling, and coordinating movement of trains would remain in-house. Private players were to participate in booking, aggregating, and delivering freight. The private partners would act as a city cargo booking office. Without any pushback from labour unions, private players were introduced. Similarly, the Minister observed that the 'core-business of the Railways is to run trains, not fry *bhajiyas*'. Trivializing non-core activities and inviting private participants required a clever placing of ideas in carefully selected words. Now, private firms are being engaged in a range of activities from modernization of stations and setting hubs for logistics to connectivity to ports.

Further, the Railways's experience revealed that introducing private firms for new manufacturing units receives no opposition, but divesting in existing factories infringes on unmanageable interest groups—the employees, management, and supply chains. In the case of the Kapurthala coach factory in Punjab, corporatization was met with stiff resistance. But five new rolling-stock factories being built as joint ventures have faced no resistance. In this regard, having a 'joint venture' is crucial because it takes the sting out of privatization. In practice, it is reassuring to the existing employees of a broader intent of the state to stay responsible and engaged, even though in day-to-day management a 26 per cent stake in the corporation makes them relatively autonomous private firms. Such inclusive development efforts require striking a fine balance between competing political and private interests and regional development needs. In sum, the Minister's left-of-centre political position and his choice of words were a part of the strategy.

While introducing change, the decision when to act is of essence. Here, for example, in the budget speech, it was a mistake to announce corporatization of the Kapurthala rail coach factory just before the assembly elections because the political climate was not conducive for change. Likewise, increasing freight charges for a benign commodity like iron ore turned out to be a political hot potato. The timing was wrong. Due to inflationary pressures, the government was facing political flak. The government asked steel producers to curtail soaring steel prices. In response, the steel producers sought a price rollback for ore freight. As a result, at the behest of the government, the Railways had to roll back the increase in freight rates. In essence, what matters is 'when' actions associated with a reform strategy are implemented.

Another aspect of the strategy was to concede battles to win the war. In both the above examples, the timing of the interventions was wrong. Despite anticipated resistance, in the democratic spirit room for consultation and negotiation with key stakeholders was built in. But such consultative process also provided ample room for an early exit. Since the unions were not confident of the benefits from change, without making it an ego issue the decisions were reversed to avoid additional trouble. Further, in the face of resistance from the affected workers, the decision to outsource toilet management at a major railway station was reversed. Finally, in the democratic context of India, an evolutionary approach to change was considered suitable. The core team argued that change creates uncertainty and raises fears among the affected. Change should be introduced in manageable quantities. Because immediate concerns dominate political discourse, an incremental approach to change was preferable, although accommodating such a range of interests amounts to funambulism, the balancing acts of a slack rope walker. An illustration of this approach was increase in axle load. Because increasing the load carried by each wagon has safety implications, and safety is a high priority in the Railways, a gradual approach was adopted. First, the load increase was tested on a non-passenger route, that too by just 2 tons per wagon. Gradually the load and the travel routes were increased.

Eventually, on some non-passenger routes, an incremental load of 8 tons per wagon was introduced. Similarly, freight and passenger tariffs were rationalized over a four-year period. When loading and unloading were increased from eight hour shifts to twenty-four hours, the change was introduced incrementally, starting in locations where unions, unorganized labour, or traders were more willing to negotiate. The objective was to introduce change where there was least resistance.

Thus political economy considerations tempered the pace and methods of introducing change. The political leadership and management strategists had to consider political positioning, choice of words, timing, and gradualism while implementing a mega transformation.

Can good economics be good politics? In the experience of the Railways the political space for populism increased with commercial success. A nearly bankrupt railway could not have financed the five rolling-stock factories that the Minister announced. But, with the successful transformation of the Railways, demand for locomotives has more than doubled, from 200 in 2004 to 500 in 2008. Therefore, building the factories was necessary and allowed for some political space in allocating them in constituencies of the Minister and other coalition partners. Similarly, the issue of employing coolies or of reducing passenger fares was absurd for a bankrupt railways, but with a Rs 25,000 crore (US$ 6 billion) cash surplus this year, such a populist intervention received no criticism, not even in the pink business and economics newspapers. With the commercial success of the Railways, the space to benefit the electoral constituency increased and so did the stature and influence of the political leadership. Yet, this success needed to be communicated to the electorate.

Initially, the media was rife with scepticism because without a track record of performance there is no positive news coverage. In the initial year the strategy was to remain relatively inactive. At the end of the first year in 2005, when the budget was presented, there had been substantive improvements, including a historic change in the tariff schedule and an incremental revenue of

Rs 5000 crore. Despite this reasonable performance, *India Today*'s 2005 annual rankings (2006) of the performance of federal ministers ranked Lalu 33, nearly at the bottom of the list. Papers like the *Business Standard* (2005) ascribed key attributes of the Railway budget for fiscal year 2006 to the Prime Minister's Office (also known as the PMO), adding that Lalu was busy campaigning in Bihar. Therefore, after the first year's performance, engaging with media required a proactive strategy. First, the ground rule was to consistently promise less and deliver more. In each budget Rs 5000 crore less was committed than was eventually delivered at the end of the fiscal year. Similarly, other measures of efficiency were stated with moderation and the results were about 5 percentage point better.

Second was to strategically select high-impact events. At the top rung was the Railway budget around which the Railways gets the undivided attention of the media—newspapers and news channels devote their prime space to discussing the popular elements of the budget. The core team seized other high-impact events too like invitations to lecture at IIMs as well as to students from Harvard and Wharton. Finally, the third bit of the strategy involved fact-based information sharing with all willing news sources without preference for any particular media outlet. In all, the media perception regarding the Minister and the Railways has been productive. In a reversal of the past ranking, *India Today* (2006) ranked Lalu second-best Minister of the year in 2006.

In conclusion, having established the political mandate and made the distinction between mandate and interference and identified the space for reform, the reformers designed ground rules for engagement in the political economy, including a media strategy. In the next few chapters, the comparative advantage of the Railways in the context of the market is analysed. In particular, the demand- and supply-side constraints and entrepreneurial responses to win back market share and margins in a fiercely competitive environment are explored.

3 The Market

'To make the cost of a phone call cheaper than that of a post card' is my dream, said Dhirubhai Ambani in 1999 (Reliance Communication 2008). At that time, people laughed off this profound statement. But Ambani was an astute businessman. He foresaw the potential market, anticipated the demand, and intercepted the future. With a discerning eye for understanding cost structures and revenue streams, the telecom industry leveraged the economies of scale to make cellular phone calls affordable for the common people of India. Now, the costs are half that of a printed post card. From 15 million customers in 1997 the telecom market has exploded to an astonishing 234 million customers in December 2007, with 8 million customers being added each month (TRAI 2008: 19). During the same period, the cost of per minute call charges dropped from Rs 17 (40 cents) to a Re 1 (2 cents) in 2007 (TRAI 2007: 2). And the private telecom firms made billions of dollars in profits.

Despite reducing passenger fares, when the Railways generated a cash surplus of Rs 25,000 crore (US$ 6 billion), why were people surprised? Here is the catch. If the Railways network were to be rebuilt today, it would need a trillion dollar investment. Once the network is laid, spreading the fixed costs over larger volumes substantially reduces unit costs because variable costs are relatively small.[1] Identifying the aspects of the business that are politically sensitive, distinguishing them from where market opportunities exist, leveraging the scope for economies of scale, deploying cutting-edge technology, and finally seizing the opportunity offered by a booming economy were essential ingredients for this success.

Prior to further discussing how the Railways was transformed, there is a need to examine the nature of the business. The railroads in India were considered to be a monopoly rail service provider (Mohan 2001b: 69 and Sondhi 2002). Yet, the same studies showed a rapidly declining market share in the cargo business (Mohan 2001b: 63). To reconcile these facts requires an under-standing of the business of railroads in India. Here, an analysis of the business is presented along with a breakdown of its revenue streams, competitive strengths, and demand elasticity of price and non-price factors. Finally, the cost structure, its variability and sensitivity to load and length of train, as well as its manipulation are discussed. In essence, the attributes that make the Railways a competitive service provider are explored in depth.

Puzzling Monopoly

What is the nature of the business of Indian Railways? If the Railways is considered a provider of rail services, it is a monopoly supplier by definition because it is the only provider of rail services in the Indian market. However, the Railways's market share[2] in 'land-based transportation of goods' eroded from about 89 per cent in 1951 to less than 40 per cent in 2004 and from 69 per cent to 20 per cent in the passenger segment, over the same period (Agarwal 2004: 175). A look at the Indian transportation market reveals to whom the market share was lost. Transportation of petroleum products shifted to pipelines. Steel and cement transportation moved to trucks and in coastal areas to shipping lines. Similarly, luxury travellers preferred air-conditioned buses and budget airlines to Railways's express trains. But the monopoly mindset and institutional pride among railway staff and policy makers hindered a hard-nosed competitive analysis, and experts continued to define Indian Railways as a monopoly, arguing for a tariff regulator (Mohan 2001b: 69). If we re-conceptualize the Railways as a transporter, then it is one among many alternative modes—trucks, ships, pipelines, airlines, and luxury buses—in a competitive market place of transportation services.

For Indian Railways this was the first step and a paradigmatic shift from thinking of itself as a monopoly service provider to recognizing the ground realities of being a transporter that faces stiff competition from alternate modes. The Railways faced the classic competitiveness problem characterized by low growth, declining market share, low productivity, poor service, and eroding margins. Solving this competitiveness problem required offering customers superior and compelling value, not mere tariff regulation. To win back the Railways's competitive edge in the market required a dispassionate analysis of competitiveness blended with insight, pragmatism, and wisdom. Here all critical assumptions about Railways's business model, variability and sensitivity of cost structures, and elasticity of demand with respect to price and non-price factors, needed a rigorous analysis.

BUSINESS PORTFOLIO

Freight transportation is central to the profitability of Indian Railways. Freight contributes 64 per cent of the Railways's traffic earnings, while passenger traffic contributes 31 per cent. The remaining 5 per cent is contributed by 'other coaching and sundry earnings'. Figure 3.1 unpacks the Railways's traffic earnings by source for the fiscal year 2004, when total earnings were Rs 42,842 crore (US$ 10 billion).

While, each day, only one-third of the 13,000 trains operated by the Railways carry freight, they account for two-thirds of total earnings. The rest is from passenger trains, the majority of which run at a loss. Railroads have a competitive edge in freight due to their ability to efficiently carry large loads over long distances. Further, much of the freight business is apolitical in nature and

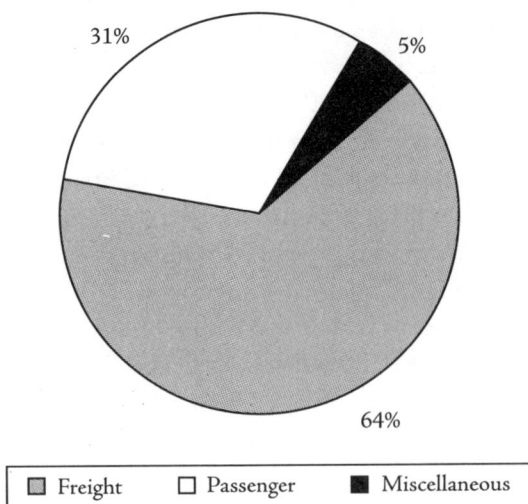

31%   5%

64%

| Freight | Passenger | Miscellaneous |

FIGURE 3.1: Where the rupee comes from

Source: Statistics and Economics Directorate, Ministry of Railways, Government of India, 2004.

can be managed on commercial principles. Thus the lynchpin of Railways's reform strategy was to optimize the freight business, reduce losses in the passenger segment, and deploy creative methods to improve the share of earnings from miscellaneous sources. But, in keeping with the nuanced approach to analysis, the reformers refrained from making generalizations at this level, and further disaggregated each market segment into its nano constituents.

Freight

In the early 1980s, because of the Gujral era decision to stop accepting piecemeal freight and focus on hauling bulk commodities, the Railways's share in piecemeal freight rapidly declined to minimal levels—except through containerized traffic and the parcel business segment. As a corollary, 90 per cent of the goods carried by the Railways consisted of eight bulk commodities—coal, iron ore, other minerals, foodgrains, petroleum products, fertilizers, iron and steel, and cement.

The expert committee attributed the declining market share, even in bulk commodities, to cross-subsidization of passenger services by freight, poor quality of services, and the national highways expansion—the golden quadrilateral and its diagonals that made the road sector more competitive (Mohan 2001a: 2). But these factors affected all commodities equally, then why was there a dichotomous response in the transportation of finished products versus raw materials? For instance, while the market share of finished products like steel and cement has been declining since the 1990s, during the same period, the share of iron ore, coal, and other minerals remained stable (see Figure 3.2 and Table 3.1).

Both iron ore and steel are heavy commodities. But, there is a distinction as well. To transport iron ore for a firm, like Tata Steel in Jamshedpur, the Railways provides a door-to-door service—from the mine pithead to the factory. The Railways picks the iron ore from Tata's mine pithead and empties it directly into the Tata Steel factory at Jamshedpur. Thus, for Tata Steel, the total logistics

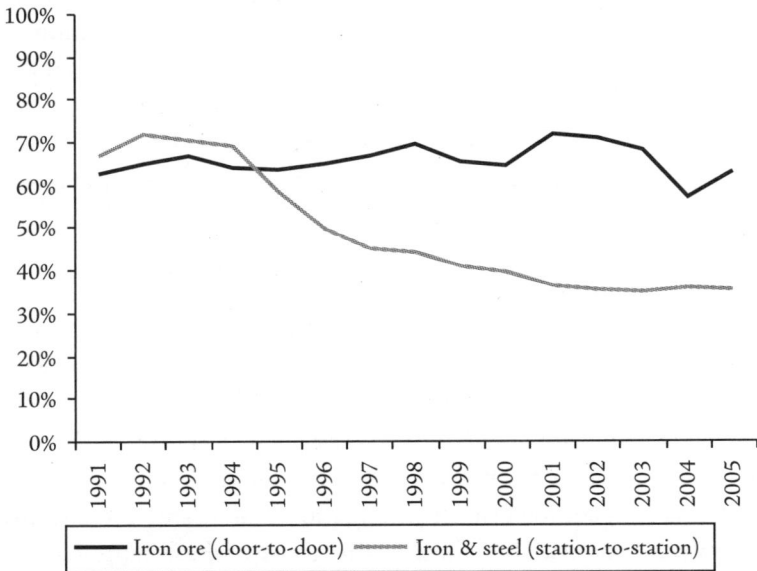

FIGURE 3.2: Market share of steel and iron ore freight (1991–2004)

Source: Statistics and Economics Directorate, Ministry of Railways, Government of India, 2008.

TABLE 3.1
Attributes of freight competitiveness

	Iron ore	Steel	Tea
Service	Door to door	Station to station	Station to station
Weight	Heavy (high density)	Heavy (high density)	Light (low density)
Quantity	More than 2000 tons	1000–2000 tons	Piecemeal
Distance	More than 100 km	More than 750 km	More than 1000 km
Overall competitiveness	High	Moderate	Negligible

Source: Authors's subjective assessment and Statistics and Economics Directorate, Ministry of Railways, Government of India.

cost is equal to the rail freight charges as other incidental costs of rail transportation are negligible. Further, rail freight charges are substantially lower than truck transport charges.

Therefore, the Railways has a competitive edge in transporting iron ore, even over such short distances as 100 km. As the distance increases, the Railways service becomes more competitive. Moreover, Tata Steel produces 5 million tons of steel annually. This production requires 8 to 10 million tons of iron ore each year or 22,000 tons every day. It would require over 2000 trucks to transport the same amount of iron ore. Transportation by road is unviable because of the number of trucks required, poor road conditions, and hilly terrain.

On the other hand, for the transportation of steel the Railways provides station-to-station service to Essar Steel, a private steel company. This steel plant has no rail sidings at the factory or at consumption centres. Further, it does not maintain any warehouses. If its factory in Hazira, Gujarat, needs to transport steel to a construction site near Mumbai, Essar has to transport the steel from the factory to the nearest railway station, and the railway hauls it to a railway station near the destination. But no steel is consumed at railway stations. Therefore, for the last leg of the freight—station to consumer—the firm once again has to organize an alternate transporter. As a result, in addition to the rail freight, incremental costs are incurred due to multiple transfers, bridging, warehouse fees, inventory, and the like. These additional costs add up as a significant component of the total logistics cost, and can exceed the cost of rail freight itself. On the other hand, road transporters provide a door-to-door service. Trucks pick up the steel from the factory and take it directly to the consumption site. Further, trucks offer flexibility for small quantities of freight, while the Railways does not accept anything less than a trainload of 2000 tons. The incremental costs of bridge transportation at both ends outweigh the cost advantage of cheaper rail freight charges. Consequently, despite steel being a heavy commodity, and rail freight charges being lower than truck, the Railways is an uncompetitive transport service provider for distances less than 750 km, and even for longer distances its competitiveness is mediocre.

To compound the problem, the Railways charged freight fares on an 'affordability principle'.[3] Cheaper commodities like iron ore were charged less, while higher-value products like steel were charged more. Such affordability-based price discrimination was benign in the Fabian era, but post liberalization Railways rapidly lost market share in transportation of steel, especially after the government revoked the 'freight-equalization' scheme for steel.[4] Based on market principles, the freight charges should have been the converse, more for iron ore that received a convenient door-to-door service and less for steel that received a troublesome station-to-station service. These flawed pricing decisions were not a result of political compulsions or broader societal considerations, but a lack of understanding about the relative competitive strength of the Railways in transportation of the two commodities.

Thus, the cross-subsidy and poor service argument put forward by the Mohan Committee and other experts failed to explain why Railways remained the preferred transporter for iron ore but not for steel. Instead, the door-to-door concept helps clarify this differential preference of consumers.

In conclusion, the Railways has a competitive advantage and pricing power in freight with the following four characteristics. First, the transportation is door-to-door (production-to-consumption), as opposed to station to station. Second, freight is in bulk quantities (greater than 2000 tons), not piecemeal. Third, transportation is of heavy goods like iron ore as opposed to low-density tea. Fourth, freight is to be transported over long hauls as opposed to short distances. If the first three conditions are met, the railway is competitive even for distances less than 500 km. However, if the first condition is not met and the Railways is a station-to-station transporter, it is uncompetitive for short and medium distances, and even for long hauls its competitive edge is diminished (see Table 3.1).

The commodities for which the Railways provides door-to-door service are raw material, ores, and minerals like coal, iron ore, gypsum and manganese, while the station-to-station service—or door-to-station—is predominantly for finished products like steel and cement. Over 63 per cent of the Railways's freight customers are provided door-to-door service, and the rest 37 per cent station-to-station service—or door-to-station, essentially involving additional logistical costs. This distinction between door-to-door and station-to-station service is summed up in Figure 3.3.

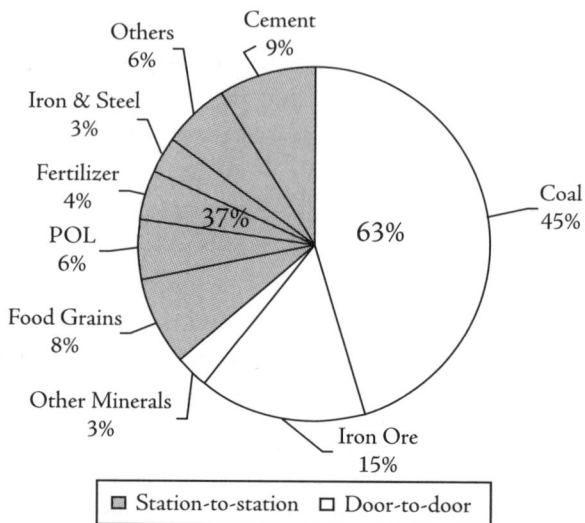

FIGURE 3.3: Composition of freight transported in 2004

Source: Statistics and Economics Directorate, Ministry of Railways, Government of India.

Several undercharged door-to-door commodities like iron ore offered ample scope for fare hikes without decreasing Railways's competitiveness in the freight market. In contrast, overcharged station-to-station commodities like steel and cement had no room for fare hikes; instead, either the fares had to be maintained or reduced along with value-added services. On a note of caution, there is no standardized policy for pricing Railways's freight services, except that there is need to have a commodity- and client-specific strategy that responds to seasonal as well as spatial variations in demand. For example, wheat and fertilizer are provided a station-to-station service, yet they offered an opportunity to raise freight charges as these services were underpriced in the past due to a misconception that freight charges for these commodities were politically sensitive.

Passenger Business

The passenger business is central to the political aspects of the transport business of the Railways. There are few business segments that offer both political and commercial flexibility. In Figure 3.4, one observes that the suburban and ordinary passenger segments constitute 88 per cent of the total number of 5.2 billion rail passengers that travel each year. Yet, they contribute only 28 per cent of passenger earnings.

This business segment that constitutes 88 per cent can be split into two. First is the suburban rail transport, essential for daily commuters in big cities. Second are slow and frequently halting ordinary passenger trains[5] often used by migrant labour, the rural poor, vendors, farmers, and office personnel to commute from surrounding districts. The Railways remain competitive in both these low-price segments, but there is a lack of political will to endorse a fare hike.

Passengers in mail and express trains constitute 12 per cent of the 5.2 billion annual travellers,[6] but contribute 72 per cent of total passenger earnings. Of the 603 million mail and express train travellers in 2004, 411 million travel in unreserved class, 155 million

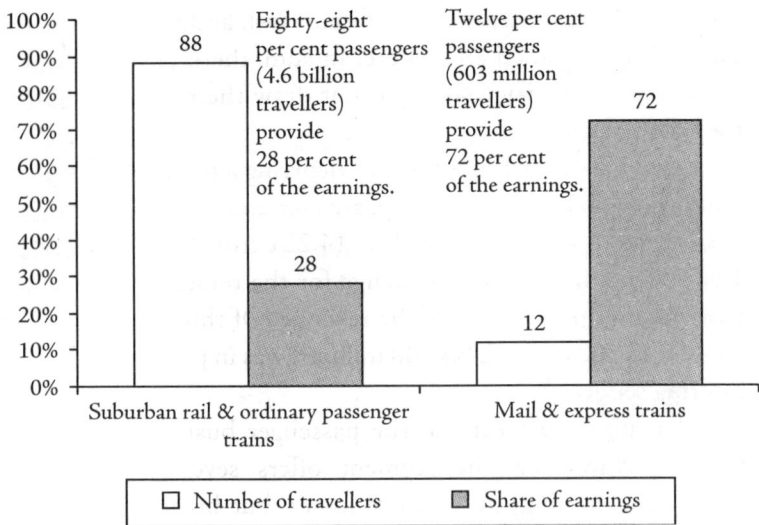

100%	Eighty-eight	Twelve per cent
90%	88	per cent passengers	passengers
80%		(4.6 billion	(603 million
70%		travellers)	travellers)	72
...		provide	provide

Eighty-eight per cent passengers (4.6 billion travellers) provide 28 per cent of the earnings.

Twelve per cent passengers (603 million travellers) provide 72 per cent of the earnings.

Suburban rail & ordinary passenger trains: 88, 28
Mail & express trains: 12, 72

☐ Number of travellers   ▨ Share of earnings

FIGURE 3.4: Number of passengers and share of earnings

Source: Statistics and Economics Directorate, Ministry of Railways, Government of India.

travel in non-air-conditioned sleeper class, and 38 million travel in air-conditioned classes. Unreserved class and sleeper class travellers constitute over 93 per cent of the 603 million mail and express passengers. These segments also offer commercial opportunities because alternate modes are dearer, but politicians are reluctant to increase fares due to the associated political fall-out.

The 38 million air-conditioned class travellers constitute less than 1 per cent of the total number of travellers, but contribute a whopping 20 per cent of the total passenger earnings. This is the apolitical segment of the high-end travel classes. However, in the past, the Railways's affordability-based differential pricing strategy taxed these passengers by increasing their fares. As a result, with the onslaught of budget airlines, the long-distance passengers opted to fly, while the short-distance travellers took to luxury buses, and taxis. But the railway staff was in a state of denial. They argued that while airlines carry 37 million passengers each year, the rail traffic was a whopping 14 million travellers each day. But the competition with budget airlines was in the small first and second class air-conditioned segment of rail travel, and the Railways was rapidly losing market share here. In sum, there was a sea of data, but either the policymakers did not draw the right inferences or they did not act.

As of 2004, the Railways was incurring a total loss of Rs 5780 crore (US$ 1.3 billion) in its passenger and other coaching business segments on an income of Rs 14,221 crore (US$ 3.1 billion).[7] This posed an enormous challenge for the reformers because the loss was over 40 per cent of the revenue. Of this loss, 20 per cent, about Rs 1200 crore (US$ 280 million) was in parcel, luggage, and catering services.[8]

Overall, 99 per cent of the passenger business is politically sensitive. However, the segment offers several opportunities to increase earnings and reduce losses, but breaking out of the passenger-fare mould required some creativity. For instance, the yield per train depends not just on passenger fare, but also on a host of other non-fare variables, like the number of coaches in a train, the combination of coaches that form a train, occupancy rates of a coach, and number of seats in a coach (that in turn depends on coach layout). In this regard, some startling results presented in Table 3.2 question the conventional wisdom that higher passenger fares invariably result in higher earnings.

TABLE 3.2
Composition of earnings per coach kilometre by
travel class for 2003

Travel class	Relative fare	Earning/coach km (Rs)	Cost/coach km (Rs)
Suburban[9]	1	22	34
Unreserved ordinary	1	22	41
Mail and express unreserved	1.82	20	26
Mail and express reserved sleeper second class	2.91	16	24
Air-conditioned three-tier	8.19	40	29
Air-conditioned two-tier	13.1	37	28
Air-conditioned first class	25.5	37	26

Source: Statistics and Economics Directorate, Ministry of Railways, Government of India, 2008.

In Table 3.2 it is worth noting that despite twenty-six times higher fares, earnings per coach kilometre from a first class air-conditioned coach are less than double those from a suburban coach. This is because the suburban rail carries 300 passengers in a coach, the air-conditioned first class has room for only eighteen travellers. Moreover, counter-intuitively, higher fares can, at times, result in lower earnings per coach kilometre. Consider the second class mail and express reserved-sleeper second class fare. This fare is about three times the ordinary unreserved fare (see Table 3.2). Yet the earnings per coach kilometre from the mail and express reserved-sleeper second-class are Rs 16 which is 36 per cent less than the ordinary unreserved coach earning of Rs 22. The yield per unreserved coach is greater because it has seating capacity for ninety passengers (and in practice accommodates more travellers) as opposed to seventy-two in the reserved-sleeper second class coach. Once more, the difference is due to the number of passengers occupying a coach. In sum, the earning per coach kilometre is not only a function of fare per passenger, but also the number of passengers travelling in a coach.

The popular conception of the perpetual and excessive demand for trains is derived from stereotypical images of thousands of passengers riding on the rooftops of trains in India, clinging to every ledge, bar, bolt, and crevice on the engine, and between carriages. But in practice, many trains are not so popular. The occupancy rate for trains not only varies by type of train and among class of travel, but also by seasons. Occupancy rates matter because an empty seat is a lost opportunity in passenger fare revenue. For every 1 per cent increase in occupancy rates, Indian Railways earns an additional Rs 100 crore (US$ 23 million). Trains heading towards hill stations are popular during the summers. But various trains that head towards the desert regions of India have very low occupancy in this period; instead it is hard to find vacant seats during the winters—the tourist season in Rajasthan. Thus, demand modelling to maximize occupancy rates on all trains across travel classes and seasons holds immense potential to enhance the Railways's earnings from the passenger business. Likewise, if an air-conditioned first class coach with a 20 per cent occupancy rate in an unpopular train is added to a popular one with a waiting list, the occupancy will increase.

The combination of coaches that constitute a train is yet another variable affecting yield per train. Consider a typical coach composition of a Rajdhani Express train. Of the seventeen coaches, two coaches provide no fare revenues while the other two offer little revenue. These include two pantry cars and two power cars cum brake vans. The remaining thirteen coaches include seven three-tier, five two-tier, and two first-class coaches. These three air-conditioned travel-classes have varying degrees of profitability—three-tier being the most profitable in practice. Thus, the profitability of a train can be improved by manipulating the combination in which these coaches are added and subtracted.

Further, the layout of coaches affects the yield per coach because an arrangement that accommodates more passengers increases per coach kilometre earnings. If the profitable air-conditioned three-tier coach layout is reorganized to accommodate eighty seats instead of sixty-four, its profitability will improve.[10] Moreover, Indian Railways predominantly has broad gauge tracks that offer opportunities to increase the seating capacity by leveraging the maximum moving dimensions.

Another variable affecting profitability is the number of coaches in a train—the length of the train. Consider Figure 3.5 where all other variables are kept constant—fare, coach class, occupancy rates, distance travelled—and coaches are added. These calculations are for mail and express types of trains that travel a distance of 1385 km.[11]

The train with sixteen coaches runs at a loss. As coaches are added to the train it breaks even at twenty coaches (Figure 3.5). Without increasing passenger fares, this train becomes profitable as the length of the train is increased to twenty-four coaches. This is because earnings increase in the same proportion as the incremental coaches, provided there is full occupancy. Yet costs associated with additional coaches are much lower. Irrespective of the number of coaches in a train, several costs remain the same—

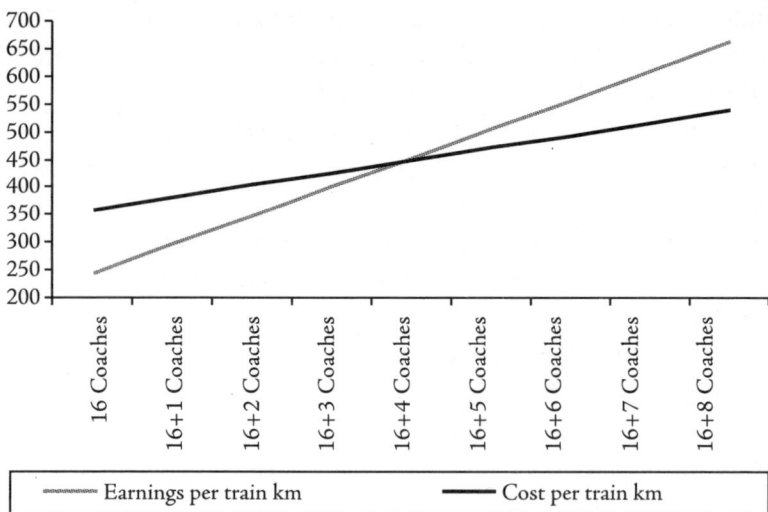

FIGURE 3.5: Train lengths and their effect on profitability

rail-track, locomotive, guard, driver, platform space, and so on. Moreover, demand for several passenger trains far exceeds supply as reflected in long waiting lists and jam-packed compartments. Keeping all other factors constant, adding coaches to a train offers opportunities to increase the yield per train as opposed to the politically infeasible task of increasing passenger fares. This offered a win–win solution.

It would thus be seen that passenger fare is only one among many variables affecting the profitability of trains. Except for passenger fare, all variables are apolitical, and have a huge potential especially if they are manipulated together to maximize profitability.

One such untapped potential for the Railways is the air-conditioned travel segment of popular trains. Many of the 155 million non-air-conditioned sleeper class passengers that travel in mail and express trains aspire to travel in air-conditioned coaches. In this regard, Lalu often narrates the experience of his friends who would try to stand near the doors of air-conditioned compartments at railway stations. This was to enjoy the little cool air that would ooze out while they peaked into catch a glimpse of what the coach looked like inside, because all air-conditioned coaches had tinted window glasses. The police would chase these curious farmers away, and this was a rather humiliating experience. This is a distinct indication of an unmet demand that presents a significant opportunity to expand the air-conditioned travel segment sixfolds, from a paltry 38 million to include the 155 million sleeper class passengers, with the caveat that fares need to be affordable. Quite like Tata's Nano,[12] the Rs 1 lakh car, or for that matter the mobile phones that are cheap enough for the common people—from vegetable vendors and carpenters to small farmers.

Miscellany

The footfalls and eyeballs of 14 million passengers who travel on trains each day, offer great opportunities for increasing non-passenger fare income in 'sundry earnings' and through brake vans and parcel vans for improving 'other coaching earnings'. Sundry earnings include licence fee for renting of advertisement space, parking, catering, and land lease. And coaching services include parcel and luggage transportation services, special trains for the use of the armed forces, and pilgrimages. But why had the share of miscellaneous earnings declined from 8 per cent of total passenger earnings in 1951 to 4 per cent in 2004? The Airport Authority of India has less than 100 airports but earns a few hundred crore rupees through revenue from advertisements. In contrast, the Railways had a paltry income of Rs 39 crore from its 7000 railways stations across India, operating 9000 passenger train services each day, in 2004. Even the metro rail transport in Singapore, a city state, had more advertisement revenue than all of Indian Railways. Cineplex chains in India have mastered the art of non-ticket revenues through catering, parking fees, auxiliary entertainment, and the like. On the contrary, Railways's land leases yielded a paltry Rs 116 crore in 2004, and the total annual catering earning of the Railways was Rs 29 crore.[13] Likewise, the Railways had a host of untapped opportunities in this business segment. What is worse, 20 per cent[14] of passenger losses were from parcel and catering services despite the fact that the Railways did not sell subsidized meals, nor did it carry parcels at a discount. Then, why were its catering and parcel earnings so low at Rs 620 crore (US$ 144 million) and losses so high?

The primary reason is that Indian Railways had acquired an institutional disinterest in commercial policies and non-passenger fare earnings. This was not on account of political interference or compulsions, but due to a lack of commercial orientation and profit motive. Moreover, the entire emphasis of the management was on operating trains. This is reflected in the preferences for posting of senior officers that manage both the commercial and operational activities of the Railways. The most coveted jobs for these officers were in operations. Issues such as catering, advertisements, parking, and licences for other commercial activities on platforms received little attention, if at all. This is evident in the fact that for years on end, licence fees had not been revised and arrears were not collected. Additionally, there was little effort towards price discovery through competitive bidding for licences and the like.

Furthermore, the commercial policy focused on retailing rather than wholesale outsourcing of non-core functions like catering, parcel, and luggage services. Each signage, parcel, was individually contracted. This led to gross underutilization of assets. For example, every train has two brake vans,[15] each with the capacity to carry 8 tons of cargo. An empty van is forgone revenue, yet less than 30 per cent of the total parcel capacity was being utilized by the Railways. Additionally, underutilization was due to a uniform pricing policy. To compete effectively in the parcel and courier market, time-bound delivery is of essence.[16] But the Railways's uniform pricing did not reflect demand for the service. Its parcel service was priced the same in the peak and lean seasons. While the parcel rates differed between fast and slow trains,[17] the quality of service varies among the mail and express trains— some are more punctual and quicker than others. While the quality of service provided by the Toofan express and Purva express varies significantly, they are both fast trains and usually tend to have the same parcel charges. Further, the pricing policy charged the same rates for transportation from the production to consumption centres, like Delhi to Guwahati, and back. While Delhi to Guwahati is in the loaded-flow direction, for which there is much greater demand, the return trip is in the empty-flow direction.

Finally, the Railways did not leverage its brand value in catering, parcel, courier, advertisement, or in its iconic trains like Rajdhani and Shatabdi. Branding offers the win–win proposition of increasing revenue several folds while enhancing customer service and reinventing the rail travel experience.

COST STRUCTURES

Most costs in Indian Railways are beyond the control of the management. They are determined exogenously. Employee salary and benefits are determined by the federal government's pay commission, diesel prices by a combination of international crude prices and government subsidies, and general inflation by market conditions. Moreover, as seen in earlier chapters, reducing the number of employees, selling or closing loss-making business segments is politically infeasible. Thus, unlike private firms, the Railways is neither able to cut down total costs nor determine the pace of the cost increase. Under such constraints, the potential to transform the Railways's finances appears slim, but in practice how formidable was this challenge?

The unit costs[18] of freight have been continuously declining at real (constant) prices since 1991 due to gains in productivity and operational efficiency. The unit costs had declined by more than half from 10 to 4.3 paise in 2004.[19] This happened due to a combination of leveraging technology and scale. With such long-term trends in the Railways, why was it heading towards bankruptcy in 2001? While the Railways's unit cost was declining in real terms, in real life operational expenses are incurred at nominal (current) prices. As the unit cost was increasing at nominal (current) prices and since corresponding increase in tariff was politically infeasible, the Railways's financial condition was deteriorating.[20]

In this regard the Mohan Committee had emphasized, 'Rate of growth in revenues has been outstripped by the rate of increase in costs' (2001a: 4). But this relationship between wage hike and productivity, expenses and revenue, can be inversed, requiring the increase in annual labour productivity to exceed the annual wage hike.

Railroads are network infrastructure that require lumpy initial investment with returns in the long term. Bigger is better because with an increase in the amount of load transported, the average cost of each ton transported falls as the fixed costs can be spread over more tons. Therefore, marginal costs are substantially lower than average cost of operations. The railroads embody strong economies of scale.

As observed in Figure 3.6 and Table 3.3, between 1983 and 2004 the gross ton kilometre,[21] increased over two times—from 518 billion in 1983 to 1176 billion in 2004. During the same period, at nominal (current) prices, operating expenses increased ten times, but in real terms—adjusted for inflation[22] to the base year 1983—the expenses have marginally decreased. To understand why the total costs do not vary much with substantial increase in the railway throughput,[23] consider Figure 3.7 that provides the composition of operating expenses.

The salary and pension expenses (49 per cent) and financial and amortization charges (14 per cent) for lease charges and depreciation are fixed. It seems intuitive that more trains imply more engines, drivers, guards, wagons, among others. However, over the years the number of trains has grown, but the number of

TABLE 3.3
Operating expenses and gross ton kilometres

	1983	2004
Gross ton kilometre (billion)	518	1176
Operating expenses at nominal prices (crore rupees)	3900	39,482
Operating expenses at real prices (crore rupees)	3400	3323

Source: Statistics and Economics Directorate, Ministry of Railways, Government of India.

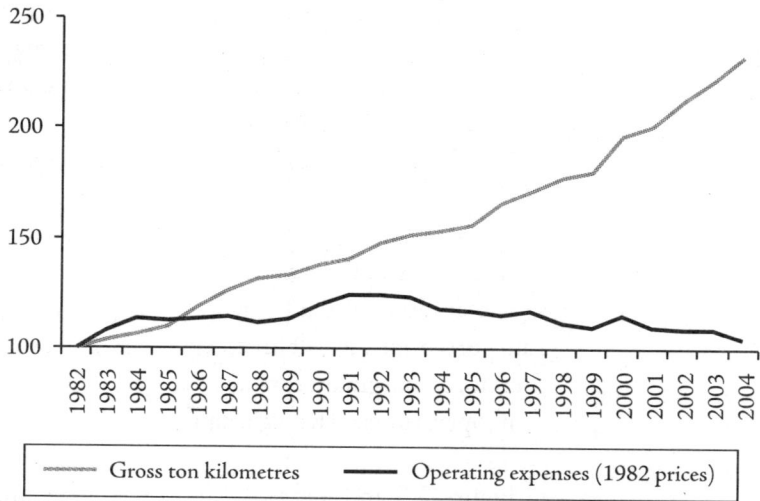

FIGURE 3.6: Operating expenses and gross ton kilometres

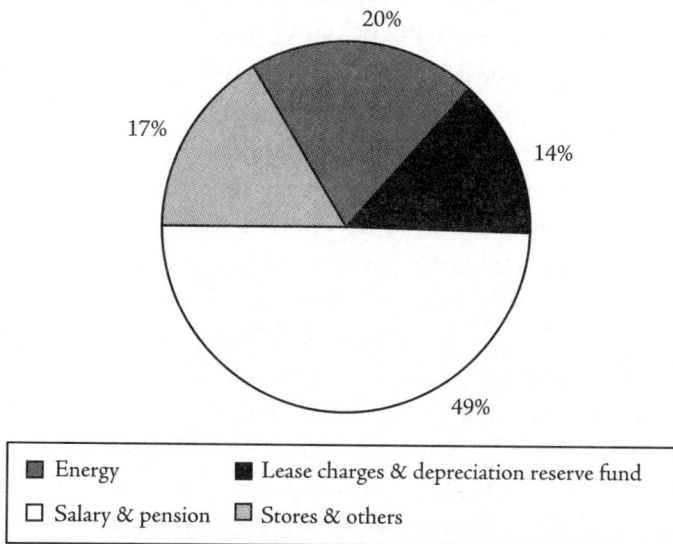

Legend:
- ■ Energy
- ■ Lease charges & depreciation reserve fund
- □ Salary & pension
- ■ Stores & others

FIGURE 3.7: Composition of operating expenses

staff, engines, and wagons, all have declined. This is because Indian Railways is a mega enterprise with enormous slack in the system. By leveraging scale, technology, operating strategy, and the like, the same assets have provided more output.

What is more astounding is that despite increase in through-put and in the number of trains, the energy and store expense do not vary much. It is intuitive to assume that more trains imply a proportional increase in diesel consumption. But in the case of the Railways this assumption falls apart. Consider the consumption of diesel over the eight-year period, 2000 to 2007. While the diesel GTKM[24] has increased 37 per cent, the consumption of diesel has increased by a third, that is, just 12 per cent.[25] Thus, costs do not vary as much with increase in throughput. The operating costs are on the rise not because of an increase in the number of trains but because of increase in the price of diesel, employee bonuses, and salaries. In essence, the Railways has low variable costs and fixed costs dominate—a trillion dollars is required to build a network as vast as Indian Railways. This is also referred to as a high operating leverage, implying that profits are elastic with respect to output or sales.[26] To put it simply, given the rigidity in the cost structure, as output increases, the operating profits increase at a rate much faster than the growth in volumes, and thus the more the freight transported, the more profitable the Railways is and vice-versa. If the Indian Railways system with its 4000 freight trains completes the to-and-fro journey faster, decreasing the time between succes-sive loadings, the Railways's output can increase exponentially. The turnaround time of freight trains was seven days in 2004. If this time can be reduced to five days then the Railways can load an additional 230 trains each day. Yet, given the high operating leverage, the increase in associated costs would be nominal and this offers a multi-billion dollar opportunity.

Further, another crucial attribute of the Railways's costs is its insensitivity to the length of the train. As the length of a train is increased—by adding more coaches—the passenger capacity increases in the same proportion. But the incremental cost is much smaller. Therefore, the unit cost per passenger declines. Table 3.4 shows the difference in unit cost of a typical Rajdhani train with seventeen coaches and a longer Rajdhani train as well as that of the case study—modelled on the Garib Rath, a low-cost air-conditioned passenger train service introduced by the Railways.

TABLE 3.4
Cost per passenger kilometre, 2008

Train		17 Coaches	20 Coaches	24 Coaches[27]
Rajdhani	Cost per train kilometre	Rs 649.20	Rs 749.03	Rs 882.55 (+36%)
	Revenue per train kilometre	Rs 993.44	Rs 1226.60	Rs 1537.93 (+55%)
	Cost per passenger kilometre	80 paise	73 paise	68 paise (−15%)
	Number of passengers per train	816 people	1032	1302 people (+60%)
Case study	Cost per train kilometre	Rs 465.48	Rs 533.25	Rs 623.92 (+34%)
	Revenue per train kilometre	Rs 567.29	Rs 686.26	Rs 851.02 (+50%)
	Cost per passenger kilometre	38 paise	35 paise	32 paise (−16%)
	Number of passengers per train	1233 people	1512	1920 people (+56%)

In the Rajdhani, as the number of coaches increase, from seventeen to twenty four, the cost of a passenger kilometre decreases by 15 per cent—from 80 to 68 paise. This is because cost per train kilometre increases by 36 per cent but the carrying capacity of the train increases at a greater rate of 60 per cent. Therefore, unit cost per passenger falls by 15 per cent.[28] This is because the longer train continues to use the same engine, tracks, driver, and crew. In essence, longer trains have lower unit cost per traveller.

Another notable feature is the insensitivity of cost to the load hauled by a train. While freight trains carry five times the load of a passenger train, costs are not five times higher, instead they are less than double the cost of the passenger train because both the trains use the same driver, tracks, signalling system, and so forth. Additionally, energy efficiency in terms of fuel consumption per thousand gross ton km of an engine hauling a heavier train is superior compared to that of a lighter train. Therefore, per ton kilometre costs of a freight train is 51 paise while that of a passenger train is almost fourteen times more at Rs 7.[29] Table 3.5 summarizes some of these attributes.

TABLE 3.5
Insensitivity of costs to load (2004)

	Gross train load	Cost per train kilometre (in Rs)	Cost per coach/ wagon kilometre (in Rs)	Cost per net ton kilometre
Passenger	1000 ton	412	36	697 paise
Freight	4000 ton	714	17	51 paise

The passenger train is not comparable with the freight train on account of several factors. But this comparison demonstrates the relative insensitivity of cost to payload. Additionally, this is borne out by an analysis of two freight trains, carrying different loads. The unit cost of a train with a heavier payload (5600 tons) is 10 per cent lower than the unit cost of a train with a relatively lighter payload (4000 tons). Moreover (see Table 3.4), a relatively lighter Rajdhani with seventeen coaches carrying 816 passengers has a unit cost of 80 paise, while a heavier and longer case-study train[30] with 1920 passengers has a much lower unit cost of 32 paise. In essence, heavier and longer trains have lower unit costs and are thus more profitable. Given the insensitivity of cost to load and length, increasing the load and length of freight trains or the number of travellers per coach and number of coaches in a passenger train offers a colossal opportunity. Alan Greenspan, retired Chairman of the Federal Reserve Bank of America, while speaking to the Federal Open Market Committee argued that a 'combination of rising capital efficiency and falling nominal unit labor costs' has been fuelling world economic growth in an era of uniquely low global inflation and interest rates (Greenspan 2007: 379). What can the Railways learn from this observation? At macro level, given the rigidity in the cost structures, if rate of growth in productivity outpaces the rate of inflation, then the unit cost of operations at nominal (current) prices is likely to fall. During the period 1991–2004, the Railways's input cost grew at 11 per cent while its volumes and asset productivity grew at 5 per cent. During the same period, its freight unit cost increased at 6 per cent per annum at nominal (current) prices. Although it would be inappropriate to assume a simple mathematical relation between the rates

(11 – 5 = 6 per cent), it is plausible to hypothesize that costs for the Railways are rising predominantly on account of inflationary pressures. In this context, the macroeconomic stability of the 2000s characterized by a benign inflation rate of 4 to 5 per cent provided a golden opportunity: if volume growth were to exceed inflation rates, the unit cost was likely to decline at current prices, making the Railways profitable.

The narrow definition of the railroad as a monopoly missed out on the fact that it is operating in the fiercely competitive business of transportation. A monopolistic state of mind was doing much harm because it called for increasing prices in the segments that were apolitical but had no commercial space, further eroding the Railways's competitiveness. Rapid sustained economic growth in India and the upswing in the commodity cycle offered mega opportunities in the transportation business. Aversion to profitability and a lack of commercial orientation are incompatible with achieving institutional sustainability, and this frame of mind needed to change so as to face the threat from severe competition in the market place. For delivering a superior and compelling value, the Railways needed to manage both supply- and demand-side constraints and reorient its operations to be more market driven and address customer needs.

An important finding from this analysis is that the profitability of railroads is not simply a function of how much cross-subsidy the freight segment provides to the passengers, or air-conditioned class passengers provided to unreserved passenger segments. Rather, profitability is a function of several variables over which the Railways's functioning can be optimized.[31] Railway expert groups and consultant reports attributed Indian Railways's high fares to cross-subsidies between the Railways's commercial and social obligations. In contrast, the reformers discovered that the Railways's fare, in the absence of cross-subsidies and in comparison with railways around the world, was very high. This is because, in purchasing power parity terms, Indian Railways's unit costs were three times that of the Brazilian or Chinese Railways. Thus, the fundamental problem was the high costs of operations and not cross-subsidies. The analysis of cost structures reveal that given the rigidity in the cost structures, the Railways can increase its profitability if it carries more volumes. Bigger is better. To carry more volumes the trains need to run faster, longer, and heavier. And to achieve operational synergy the Railways requires a systemic approach. Such a coordinated approach is at the core of the supply-side constraints, if the unmet demand is to be addressed. How these insights were translated into billions of dollars will be discussed in the following chapters.

4 Milking the Cow

The Minister was discussing the freight-loading target for the forthcoming budget for fiscal year 2004–05. In 2004, the Railways had achieved a freight loading of 557 million tons. The Railway Board was of the opinion that even a modest increase at a meagre growth rate of 2.3 per cent from the previous year to a target of 570 million tons would pose a challenge. But such a conservative proposal was not unusual because the compounded annual growth rate of freight over 1992 to 1999 was 1.86 per cent (Mohan 2001b: 16). The Board members articulated their concern, 'The railway assets are over utilized. On some routes trains are running at 140 per cent of capacity.'

Since the Board could not reach a consensus on the target for freight loading for fiscal year 2004–05, the Minister called for the relevant 'file'. The budgetary freight-loading target for 2005 was set at 580 million tons and an internal 'mission 600 million tons' was launched. The Minister personally wrote to all General Managers in zonal railways encouraging them to make every effort to honour the target. With the very same people, network, and rolling stock, Indian Railways not only achieved its stretched target but exceeded it by loading 602 million tons in 2005. In Indian Railways's 150-year history, this 45 million tons of incremental loading was the highest ever in one fiscal year. Since 2004, Indian Railways has graduated from 557 million tons to 795 million tons in 2008. This increment of 238 million tons in four years is significantly greater than the 155 million tons the Railways clocked over the entire 1990s. Based on the present trend, the Railways is expected to record an additional loading of 300 million[1] tons during the five-year period 2005–09. In comparison, this is more than the 280 million tons of incremental loading recorded over the previous two decades (1981–2001).

What enabled Indian Railways to make this quantum leap from 10 million tons to 60 million tons a year, and from 2 to 8 per cent annual growth rate? Simply put, this is the result of a scale-driven strategy of growth. For playing on volumes the Railways needed to manage both supply- and demand-side constraints. And the entire supply-side management can be attributed to three interventions—faster, longer, and heavier trains. Each of these three elements is worth a billion dollars in profit. Faster, longer, and heavier trains are economical. They use the same resources—locomotives and crew, routes, wagons, and systems—and transport more freight. Marginal costs are negligible, resulting in lower unit costs, but revenues increase substantially. World over efficiency is equated with heavy haul trains, led by the Rio Tinto in Western Australia. A typical Rio Tinto Pilbara train has a loaded weight of 30,000 tons and length of over 2 km (*Railway Gazette*, 19 June 2008). In contrast, Indian Railways's axle load[2] was 20.32 tons and a train with fifty-eight wagons hauled a payload of over 3364 tons.[3]

Leveraging the fact that costs are sensitive to train kilometres rather than passenger kilometres or ton kilometres, the strategy was to improve yield per train. This strategy amounted to refocusing the Railways's comparative advantage to its basics. These attributes—namely guiding (allowing faster speed), coupling (facilitates demand responsive lengths), and bearing (allowing heavy axle load)—were the motivations for inventing railroads in the mid-nineteenth century (Van der Meulen and Möller 2006: 1).

First are faster trains. Here, faster does not refer to the actual speed of trains but quicker turnaround—reduction in the time lapsed between two successive loadings. Indian Railways had 4000 freight rakes in 2004. With a seven-day turnaround time the

Railways loaded 570 trains each day. Reducing this turnaround time to five days enabled the Railways to run 800 trains a day, that is an additional 230 trains day after day. All else being constant, incremental revenue from these trains alone amounted to Rs 10,000 crore (US$ 2.3 billion).

Second is running heavier trains. In 2004, the Railways was loading 30,000 wagons each day, which increased to 40,000 in 2008. Adding an extra ton of load to each wagon translated into 40,000 additional tons or 15 million tons a year. Instead of 1 ton, 6 tons were added to a wagon, translating into 90 million tons of incremental load each year or Rs 6000 crore (US$ 1.4 billion) in incremental revenue.

Third is increasing the length of trains by attaching additional coaches in popular mail and express trains with long waiting lists. Each additional coach provided a crore in incremental revenue and 3000 such coaches would translate into Rs 3000 crore (US$ 0.7 billion) of additional revenue. Similarly, extra wagons were added to covered and open freight trains—40 to 41 in the former and 58 to 59 in the latter.

Three-fourths of the supply-side transformation of the Railways is explained by these three interventions. If these interventions were so simple and intuitive, why were they not cashed in upon earlier? While all the strategic solutions emerged from within the bureaucracy of the Railways, and are in fact as old as the inception of railroads, they did not translate into action because of the lack of a systemic approach and cross-functional and spatial coordination. Indian Railways is a mega system with several sub-systems within systems, quite like a *matryoshka*, the Russian nested doll. These systems are structured around various departments within the Railways and are interwoven and interdependent in complex ways. Each technological intervention required collaboration across departments, zones, and divisions, and to expand the capacity of the system there was a need to work on several variables simultaneously to synergize policy initiatives. To capture the complexity and the interrelationships between departments and within policies for each of the interventions consider the following.

Faster

The turnaround time of a wagon is a function of several variables. For reducing the turnaround time, either the speed of the trains had to be increased, or the non-travel time reduced, or both. In 2004, the turnaround time was seven days. Of the seven days, two were spent in travel[4] and the remaining five in non-travel activities. Among the three non-travel activities, handling—loading and unloading of goods—at terminals took two and a half days, train examination took another half day, and the residual two days were spent on miscellaneous tasks including maintenance and asset failure. It was discovered that train speed had remained almost unchanged over the last thirty years.[5] After little success with increasing the speed of trains, an aggressive multipronged strategy was launched to reduce non-travel time spent on handling, train examination, and miscellaneous tasks.

First, detention time was longer because of shorter platforms and restricted working hours at many terminals. While the Railways run trains day-and-night, at many goods sheds the handling was restricted to business hours—twelve to sixteen hours a day. The time allotted to customers for handling was nine hours beyond which a penalty was charged. Consider a typical terminal that operated between 6 am and 6 pm. A train that arrived after 9 am took nearly twenty-four hours to load, yet no penalty was applicable because the loading time had to be computed only within working hours. This problem was compounded by short loading lines and platforms—a legacy of the pre-Gujral practice of piecemeal loading. Since several platforms were less than 685 m long, trains had to be loaded in more than one instalment and each instalment took up to twenty-four hours.

Therefore, it was decided to convert all goods terminals that handled more than ten rakes a month into round-the-clock working terminals with full-train-length platforms. Adequate funds were made available to carry out these works.[6] The implementation was time bound and projects were to be completed within six months.

These initiatives were coordinated with several operating and commercial policy interventions. Preferential traffic schedule[7] was modified to grant precedence in allotment of rakes to efficient customers that fell in the same category—having full-length terminals, mechanized handling facilities, and round-the-clock working. Demurrage and wharfage charges were increased and, with a few exceptions, the discretionary powers to waive these charges were revoked. For efficient freight handling, incentives such as the engine-on-load scheme were introduced. Unlike before, now the engine would wait if loading was completed quickly—three to six hours. In essence, efficiency was rewarded, and delays penalized.

A second critical cause of delay was examination of freight trains. In a quest to reduce operational delays, Gujral had eliminated all enroute train examinations. Since then, most trains were examined prior to loading and the validity of examination expired at the end of every trip—thus known as end-to-end examination. However, this examination was independent of the distance travelled by the train, be it 5000 or 500 km. This shortcoming in the examination practices was pointed out by the Minister with his rustic humour. In a meeting with the Railway Board he quipped, '*Tumhari gadi ko toh steshan ka naam padhna aata hai. Kyonki yeh toh station ka naam padh kar bimar padti hai*' (your trains are literate; they seem to fall sick after reading the name of stations). On the other hand, for some trains, operating on a closed circuit with a home base, the validity expired after a specific number of kilometres were traversed. These trains were examined after a couple of trips.[8] Under this dualism nearly two-thirds of the trains were examined after every trip while a third were under the efficient system based on distance travelled. As a result, of the 600 trains loaded each day, over 300 trains were examined each day. Conducting an examination and delays in organizing locomotives took about fifteen hours. Therefore, each day 4500 train hours or 188 train days were spent examining trains.

As part of the faster turnaround strategy, the train examination frequency has been revised to reflect mileage and time lapse. Examinations are conducted after every 4500 km of travel or every fifteen days, whatever is earlier. This has eliminated the need to examine trains at the end of every trip. Further, the validity of the brake power certificate—a service guarantee on speed braking efficiency within a specific braking distance of the train—for closed-circuit rakes has been increased from 4500 to 7500 km, or thirty-five days, whichever is earlier.

For the third category of non-travel time spent on miscellaneous activities—maintenance, locomotive breakdowns, traction change points, and the like—constraints were mitigated. For example, there were several traction change points. As early as the 1960s the links from Delhi to Kolkata had been upgraded to electric traction. However, the feeder branch lines—the first and the last mile on the same route—were not electrified. This resulted in ten to fifteen hour delays at the traction change points in organizing a diesel locomotive, crew, and other operational requirements. In the past, the railways neither had incentives for the customers to migrate to electric traction nor did the Railways electrify the last mile. The transformation strategy identified this as a high-priority intervention and electrified the last mile at its own expense with large savings due to reduction in delays. Intervals between two consecutive repair and maintenance activities were increased and delays were curtailed. These efficiency gains were made by reducing the delays in queuing for repairs, repairing sick wagons, and their release and redeployment. This further enhanced the supply of wagons and coaches. Finally, information technology was deployed. Through a freight-operation information system real-time rake movements were monitored, resulting in efficient management of wagons. Through a combination of these initiatives,[9] deployed simultaneously, the wagon turnaround time decreased from seven to five days. Similar efforts were made to reduce the turnaround time of coaches. And rationalization of locomotive and crew linkages improved the availability of locomotives for freight trains. The mentality of a hungry tiger was of essence—a sense of urgency, sharp focus, aggressive chase, and swift execution. The entire intervention cost a few hundred crore rupees, but translated into US$3 billion (Rs 13,000 crore) in incremental revenue.

Longer

To run longer trains, with twenty-four coaches, over the 1990s the Railways procured higher horse power locomotives at three times the cost of older and slower ones. But longer passenger trains required longer platforms. However, the length of platforms was not increased to accommodate longer trains. Thus, the Railways continued to run passenger trains with an average length of fourteen coaches. In a pithy statement, the minister observed, 'Indian Railway purchased a Jersey cow, but forgot to milk it.'

Therefore platforms were lengthened on a priority basis, but here, demand followed supply. Trains were lengthened first, followed by the decision and allocation of funds to extend all platforms along the route of the train, located in multiple zones requiring inter-zonal coordination. As a result, while passengers did face inconvenience for a bit, this induced imbalance created a sense of urgency for constructing the platforms.

Longer trains required more coaches but, in the short run producing more coaches to replace the existing stock was impossible. Instead, more effective and efficient utilization of existing coaches seemed more practicable. This asset optimization was tackled through a three-pronged strategy. First, the allocation of coaches was reorganized by detaching coaches from trains with less demand and adding them to trains with long waiting lists. Second, like in the case of freight wagons discussed earlier, rationalization of examination and maintenance practices resulted in enhanced availability of coaches—brake power certificates were issued for longer distances, and intervals for routine maintenance were increased. Third, the same rake composition was run more often. While the first required responding to passenger demand on a seasonal basis, the last required standardization of coach composition, so that the same train could be utilized for multiple trips within a day. For example, consider an express train that arrives in Delhi at 9 am and departs at 6 pm. Now, either this train is renamed and set off as another express train or it makes a short-distance interim trip. Such changes required standardization of the composition of coaches among a set of interchangeable trains.

Once the coaches were available, their deployment was prioritized based on the potential yields. A passenger-profiling management system was developed to provide detailed demand (and lack of it) for each class of travel on each train. All chief commercial managers in each zonal railway started compiling a daily status report of this demand model. Profitable coaches, like air-conditioned three-tier coaches and in-demand parcel vans on long-distance trains, were given precedence. Similarly, the length of covered (BCN) and open (BOXN) freight trains was increased from forty to forty-one and from fifty-eight to fifty-nine wagons respectively. This was achieved within the permissible moving dimension—namely constraints imposed by the length of loop lines.

HEAVIER

The Minister had been receiving complaints regarding overloading. On 27 September 2004, Lalu arrived at Muri where the entire train could be weighed on an in-motion electronic weighing bridge. At the traffic control room in Rail Bhavan, the senior management coordinated a nationwide weighing of trains. Within four days (27–30 September 2004) 101 freight trains had been weighed and in most trains massive overloading—ranging between 2 and 15 tons per wagon—was found. This revealed that the railways was hauling the additional load anyway, but was not getting paid for it. This practice had been prevalent for years, but substantive damage to the tracks or rolling stock had hardly ever been reported.

Apart from disciplinary action, the reformers sought to fix the problem, not just people. By increasing the axle load—that is to say load per wagon—they significantly reduced the room for overloading. In the face of substantial scepticism, and some criticism, the Minister took a strong stand regarding the increase in axle load. In Lalu's words, 'My mother has taught me to take the bull by the horns; catching it by the tail lands you with a kick in the face.' Since safety issues are 'political hot potatoes' in the

Railways, and increasing axle load raised the safety alarm, its implementation required a thorough assessment of risk and its mitigation. This decision to take a calculated risk stemmed from the field inspection where the minister had discovered the illegal overloading and from the subsequent revelations of the extent of such practices across the Railways. Moreover, over the 1990s, Indian Railways invested US$ 6 billion in strengthening its track structure.[10] During this period the Railways commissioned three independent assessments—CANAC Incorporated (a Savage Company), experts from the Asian Development Bank, and an empirical assessment by the International Heavy Haul Association. With some improvements,[11] these agencies certified the tracks fit for hauling 25 tons axle load. Yet the Railways continued running trains based on an axle load of 20.30 tons. What makes this intriguing is that as early as 1922, in some zonal railways, on the older and weaker 90 pound rails, steam engines[12] operated on axle loads of 22.5 and 23 tons. The actual axle load was greater because of a hammer blow effect—emerging from a lack of wheel alignments (imbalances) that created greater stress on the tracks. If weaker tracks could carry heavier load why were the new and stronger tracks carrying only 20.3 tons axle load?[13] Moreover, until the 1990s electrical locomotives of class WAG4 and WCG2, with 21.9 and 22.5 tons of axle load respectively, were in operation.

To operationalize increased carrying capacity of wagons required an addition of four springs in each wagon, upgrading of tracks required improving the weaker segments that were built with the old rail, and a revision of the stress calculations with the track modulus to reflect the tensile strength of the new tracks. Further, electronic weigh bridges and wagon impact load detectors (WILDs) were installed to monitor and prevent overloading on trains. Plus higher horsepower locomotives were assigned to haul heavier trains and specific commodities were identified and reclassified to allow heavier loading of wagons. There were a host of complementary interventions. Finally, a visit was made to the safety commissioner for approval.

The Big Five

Five strategic themes emerge from the preceding discussion. First, thinking beyond the resource constraint required leveraging resources such that aspirations exceeded the resource endowment of the Railways. Here innovation and asset optimization—as opposed to asset accumulation—were central. The strategy was to fully utilize assets by running faster, longer, and heavier trains. Second was coordination and cooperation requiring functional and spatial synergy as well as complementarities among various kinds of policy interventions, such that the sum of parts was much greater than the whole. Third was to fill in the gaps with strategic investments—low hanging fruits. Low cost, short gestation, high return, and rapid payback were the criteria for these investments. Fourth was an aggressive chase and follow-up mechanism built on monitoring systems, evaluation, and system feedback, facilitating swift implementation. Fifth was a deliberative and calibrated approach where projects were first piloted to learn, revise, and scale up in a phased manner. However, overarching these themes was the organizational mission to champion inclusive reforms— the political mandate of transforming the financial condition of Indian Railways without burdening the poor travellers and railway employees.

Leveraging to Optimize

Simply put, earlier, the Railways's modest aspirations were dwarfed by its gigantic resources. Assets were underutilized and the system was slack, lumbering along like an old elephant. The extent of slack in Indian Railways is illustrated by a quick benchmarking with comparable railways around the world. All class one freight railroads in the United States[14] have about one-tenth the number of employees, but produce seven times the Indian output (measured in ton kilometres). Similarly, the Chinese Railways has the same length of network, and only marginally larger proportion of double track, as that of Indian Railways. With twice the number of locomotives and wagons, and the same passenger transportation output (measured in passenger kilometres), the Chinese Railways had five times the freight throughput of Indian Railways (UIC 2006). While these railways are not strictly comparable, these comparisons do illustrate the extent of slack in the Indian Railways system. Worse, the annual ritual of setting an output target was relaxed, if not sluggish, at 2 per cent year-on-year growth. This neither inspired action nor required efforts to achieve. While the Railways consistently met targets, it simultaneously faced a financial crisis.

In 2005, when the annual target was stretched from 2 per cent to an ambitious yet realistic 8 per cent, the options were limited. Given the temporal and fiscal constraints, adding tracks and rolling stock within the same year was practically impossible. Thus, the central emphasis of the strategy gravitated towards getting more milk from the same cow, as opposed to buying more cows. This approach to achieving stretched goals through resource leveraging is articulated by Hamel and Prahalad (1994: 172). They characterize organizations based on the combination of ability—or not—to set stretched goals and leverage resources.[15]

	Lack ability to leverage (unstrategic)	Leverage resources (strategic)
Slack goals (unambitious)	Loser: Firm with neither ambition nor ability to leverage resources.	Sleeper: Firm that has 'nascent capacity' to leverage resources but lacks ambition.
Stretched goals (ambitious)	Dreamer: Firm that thinks big but does not leverage its resources.	Winner: Firm that stretches goals and achieves them through resource leveraging.

Indian Railways graduated from a sleeper to becoming a winner by deploying a combination of setting stretched targets, leveraging resources to optimize existing assets, working through cross-functional teams, fostering strategic alliances, investing strategically, adopting a deliberate and calibrated approach, and finally chasing projects to swift completion to reap high returns.

Functional and Spatial Coordination

A prerequisite to operating faster, longer, and heavier trains was adopting a systemic approach towards asset optimization. As seen earlier, enhancing the effectiveness and efficiency of the Railways required a complex multipronged approach because of the interdependent and interwoven structure of Railways's functioning. For instance, consider decreasing the wagon turnaround time. This single target of reducing wagon turnaround time from seven to five days required coordination among decisions regarding investment, commercial, and operating policies as well as train maintenance and examination practices.

Mechanical and traffic departments had to collaborate for making train examination practices more efficient and less frequent. Mechanical, electrical, and traffic had to collaborate to improve the availability of locomotives and crew by link rationalization and decreasing outage—time spent on queuing, refurbishing in workshops, and release—and finally redeployment. Civil and finance had to coordinate in order to determine cost-effective investment for improving poor infrastructure, illumination, paving of access roads, equipment, and tools at the examination depots. Finally, the traffic department had to modify the preferential traffic schedule and commercial policies concerning demurrage and wharfage charges. In essence, mechanical, electrical, civil, finance, and traffic departments were required to work together as a cohesive team for reducing the turnaround time. Further, spatial coordination and cooperation were required, and then various zonal railways as well as divisions within zones had to cooperate and collaborate.

The complexity and interdependence for a single initiative and the inherent tensions and conflicting departmental interests are captured in the case study of the taskforce on train examination, constituted on 29 November 2004. This 'Multi-disciplinary Taskforce on Freight Train Examination Practices and Procedures' comprised the directors of the mechanical and traffic departments with the director mechanical as convener. The primary objective of the team was to 'reduce overall terminal detention for train examination' without compromising safety. After making field visits to some zonal railways and seeking inputs from all zones, the taskforce submitted its report, but the Railway Board got stuck in a stalemate. Each department was concerned with its own perspective, while the traffic department wanted train examination practices revised so as to increase availability of rakes, the mechanical department's primary concern was safety. Some key requirements and concerns of the traffic and mechanical departments are summarized in Table 4.1.

Yet there was significant progress and both departments agreed on some central issues. First, wagons are the bread-winning rolling stock of the Railways, and therefore needed to be well maintained and optimally utilized. Second, terminal detentions, both for train examination as well as loading and unloading needed reduction. Third, infrastructure facilities needed to be upgraded at maintenance and examination depots plus goods sheds.

In order to break the gridlock and reconcile the differences, the Minister issued the following memorandum to the Chairman Railway Board on 18 March 2005.

TABLE 4.1
Requirements of the traffic department

1. Increase operational flexibility and enhance availability of wagons.
2. Replace the criteria for examination from trip based to mileage based.
3. Increase validity of brake power certificates for close-circuit rakes from 4500 to 6000 km.
4. Scrap the practice of post-tippling and loading train examination.

Concerns of the mechanical department

1. Safety of trains is paramount.
2. Lack of infrastructure and diagnostic facilities at examination and maintenance depots.
3. Lack of operating discipline—close-circuit rakes get jumbled up and trains run despite expired brake power certificates.
4. Greater delays due to placement and release detentions (ten hours) rather than train examination per se (five hours).

Source: Taskforce on Train Examination Practices, Railway Board, Ministry of Railways, Government of India, 2008.

On account of difference of opinion between Traffic and Mechanical Directorates, the report of the Task Force on Train Examination has not yet been put up. CRB (Chairman Railway Board) should try to resolve the differences keeping in view the following:

(i) Train examination practices should aim at achieving the twin objectives of operational flexibility and enhanced availability of rakes without compromising safety of trains.

(ii) The opportunity cost of these kinds of excessive detentions runs into hundreds of crore of rupees.

(iii) The time spent on placement and release is two times more than the time spent on train examination per se. This needs to be brought down by *at least 50%*.

(iv) The validity of BPC (brake power certificate) in end to end rakes is decided not by quality of examination but by the distance travelled between two successive loadings. This appears not only to be incongruous but also leads to examination of rakes after travelling very short distances in number of cases. This needs to be examined and suitably addressed.

(v) We should try to reap full benefits of massive investments made in procuring superior technology rolling stock. We should also *bench mark our train examination practices with the best in the world* and make suitable investments in upgrading infrastructural facilities.

We have achieved Mission 600 MT with excellent team effort and we should commit ourselves, rising above departmental considerations, to achieve Mission 700 MT with the same team sprit. This should be put up at the earliest and in any case not later than 11.04.2005.

Even after the above Memorandum, it took over two months to resolve differences between the departments. Meanwhile, there was relentless follow-up, providing a sense of urgency. In May 2005 the Railway Board decided to introduce a comprehensive premium end-to-end service wherein rakes were examined, not at the end of each trip but at twelve day intervals. Decisions emerged through patient deliberation, diligent follow-up, and respect for mutual views in a democratic frame of mind, with the spirit of accommodation as well as calibration leading to consensus and eventually to action. Funds were allocated for upgrading infrastructure facilities and authority to procure spare parts was devolved to the field units.

Similar taskforces were constituted for increasing axle load, passenger amenities, leveraging information technology for productivity gains, tariff rationalization, freight incentive schemes, redesigning wagons and coaches, route-wise planning for high density networks, public–private partnership, and so on. And, in practice several initiatives were pursued all at once, making the whole engagement a super-intricate process.

The cross-functional teams were constituted with members drawn from relevant departments. Each team had a convener appointed from a lead department whose expertise was directly aligned to the team's objectives. For example, the team on axle load was convened by a civil engineering officer, tariff rationalization by a traffic officer, and train examination practices by a mechanical officer. These taskforces were not new to the Railways. However, the reformers upgraded these teams from being merely deliberative bodies to decision-making bodies. The advisory reports produced by these teams were placed at the centre of decision making and results were sought in a time-bound manner. The teams pushed the Railways to question entrenched practices and initiate deep-rooted change. The experience of the Railways with cross-functional teams as a critical vehicle for leading change is not unique. Similar experiences are shared by corporations as well.

The cross-functional teams lie at the heart of what people call my method. They were the key to the success of the Nissan Revival Plan, because they necessarily engaged those who would be charged with carrying out the plan.

I knew that if I tried to impose change from the top down, I'd fail. That's why I decided to place a battery of cross-functional teams, or CFTs, at the center of the recovery effort. I'd used CFTS on the other occasions when I was working to turn a company around, and I'd come to the conclusion that they were an extremely powerful tool for inducing executives to look beyond the functional and spatial boundaries of their direct responsibilities. The idea was to tear down the walls, whether visible or invisible, that reduce a collective enterprise to a congregation of groups and tribes, each with their own language, their own values, their own interests [Ghosn and Ries 2005: 102–3].

However, the distinction lies in the fact that while the Railways is a mega-governmental organization, essentially a Ministry in the Government of India, corporations like Nissan, provided their management with the flexibility associated with private corporations.

Strategic Investments

Various cross-functional teams on rolling stock identified and prioritized investments that were vital for ensuring effective utilization of the stock. The recommended inputs included increasing the length of the platform and goods terminals, upgrading and strengthening infrastructure at examination and maintenance depots, and deploying information technology for strengthening the freight and passenger-operating information systems.

All such investments are strategic in nature—short gestation (less than a year), low cost (few million dollars), rapid payback (within a quarter), and high return (ten to hundred times). By investing a few hundred crore rupees (few million dollars) for increasing the lengths of the goods sheds and passenger platforms, illuminating and improving the access roads to goods sheds for round-the-clock working, and upgrading and strengthening maintenance and examination depots, the Railways raked in manifold returns in the form of incremental income—US$ 1 billion (Rs 4300 crore) in 2006 to US$ 3 billion (Rs 13,000 crore) in 2008.

Likewise, the taskforce on throughput enhancement suggested elimination of critical bottlenecks on high-density networks and congested junctions for augmenting network capacity. All this was achieved through route-wise planning simulations deploying information technology for optimization of the system. The focus was on increasing the overall capacity of critical routes by enhancing throughput per train or augmenting the number of trains on a particular route, or both. Consider the following four illustrations. First, as recently as 2004, the high-density Delhi–Mumbai and Delhi–Kolkata routes—each stretching over 1400 km—had less than 100 km of track built of the weaker older rail, requiring the trains on these routes to function once again on the lowest-common denominator of 20.3 tons axle load. The entire investment did not yield results as axle load could not be increased due to these weak track segments. When prioritized investments were made for upgrading these few miles of tracks, they yielded enormous returns because they unlocked billions of dollars in prior investment on track improvements. Second were low-cost traffic facility works, namely improved signalling systems and so forth. For instance, on all high-density routes, intermediate block signal systems are being installed at a cost of Rs 250 crore. As a result the route capacity will improve by about 10 per cent. Third, was time-bound completion of ongoing doubling[16] projects and other throughput-enhancing last mile projects on high-density networks. Fourth, steps for decongesting busy junctions by constructing bypasses, underpasses, flyovers, and crossovers, ameliorating constraints at goods sheds and coaching terminals were initiated.

In essence, the present strategy made a clean departure from the past routine. Earlier the emphasis was on acquiring new rolling stock and building tracks as opposed to maintaining the existing— half a trillion dollar worth—of capital stock[17] and utilizing it effectively. This obsession with allocation, expenditure, procurement, and construction is not solely a public sector malady. Hamel and Prahalad (1994) articulate this concern in the context of private corporations—IBM, General Motors, and Philips—as well:

[T]he resource allocation task of top management has received too much attention when compared to the task of resource leverage. …[T]here has been relatively little emphasis put on top management's role in accumulating and orchestrating a firm's resources. …[W]hatever the efficiency of resource allocation, sooner or later, in every industry, the battle revolves around the capacity to leverage resources rather than the capacity to outspend rivals [p. 174].

In contrast, the primary focus of this investment strategy has been to get more from the existing assets by effective and efficient utilization. This has been achieved through decongesting the network, reducing transit time, and enhancing the utilization of rolling stock to increase the throughput of traffic (Budget Speech 26 February 2008: 3). Earlier there was an overall liquidity constraint and the strategic investments mentioned were not granted adequate priority and resources. The reformers ensured that there were no liquidity constraints for such investments and these initiatives were accorded high priority. Initially, precedence was granted to low-cost projects and as Railways's surplus increased, liquidity constraints were relaxed for all throughput-enhancement projects. Additionally, instead of a once-a-year ritual, such projects are now sanctioned throughout the year, further facilitating timely improvements. To allow swift action, authority to sanction small works, award contracts, and finalize estimates was devolved to the front line of action—Divisional Railway Managers and General Managers have been freed of the need to seek approvals from the Railway Board.

Fostering Strategic Alliances

To meet soaring wagon demand, strategic alliances were forged. Customers were invited to lend a helping hand. Through the opening up of container trains to private players—other than railway-owned container corporation (CONCOR)—investment worth another Rs 1000 crore poured in. Since February 2006, when these concession agreements were signed, over a period of two years, about fifteen new private firms have entered the market and added seventy new container trains—this is half of all the container trains added over the last twenty years by CONCOR. Moreover, several inland container depots that act as dry docks are under construction of which two have been built and are in operation and another five will resume operation this year, while a dozen are in various stages of completion.

Through a wagon investment scheme that guaranteed availability of rakes to wagon owners, Rs 3000 crore worth of rolling stock was leveraged from the market. Not only did this augment supply but it also unleashed innovation where the private sector is experimenting with the production of more efficient models of wagons. The role of the Railways's Research Design and Standard's Organization (RDSO) has evolved from that of the sole designer and certifier of standards to that of assessing and monitoring quality and safety. Additional schemes, like a matching grant from the Railways for construction of private railway sidings and public–private partnerships for port connectivity—in some cases build-operate-and-transfer arrangements—were contracted.

Through a strategic alliance with ancillary units, the Railways ramped up production from a 100 to 250 locomotives a year for both diesel and electric engines. In-house production focused on core strengths—only those production aspects that could not be procured from subcontractors in ancillary units. This outsourcing was popular among management and labour unions alike because it was done only after maximizing the number of shifts for in-house production, as well as freeing up overtime rules to allow full employment and complete participation of existing labour. Further, production cost per unit declined as the same staff and overheads were distributed over more output. In essence, the focus of strategic arrangements was to foster win–win alliances where in-house capabilities were insufficient to meet supply-side constraints.

Deliberative and Calibrated Approach

A first step towards reducing train examination time was reducing frequency of examination. Progress on this initiative was calibrated and deliberative as seen in Table 4.2.

Similarly, the reformers adopted a calibrated and incremental approach towards increasing axle load. Initially, heavier trains were run on non-passenger routes, with considerations given to safety, track structure, commodity hauled, and potential to scale the operations for a sub-set of high-demand commodities like iron ore and coal. A more complex example of this approach is presented in Appendix 4 where 45 rate circulars issued over four years testify to the gradual increase in the axle load on selective routes for some bulk commodities.

The deliberative and calibrated approach focused on piloting initiatives to learn, revise, and finally scale up. This approach was adopted for several reasons. First, making changes in small increments faced lower resistance in a cautious and at times

TABLE 4.2
Calibrated approach to reducing train examination time

Timeline	Incremental change in examination practice
October 2004	6000 km or 20-day close-circuit rakes introduced at 'A' category closed-circuit bases.
May 2005	Premium end-to-end rakes with 10 + 2 days validity introduced for BCN wagons.
February 2006	7500 km or 30-day close-circuit rakes introduced at select 7 close-circuit bases.
April 2006	Premium end-to-end rakes with validity of 12 + 3 days introduced at selected 40 examination points. Premium pattern introduced for other wagons also like BOXN, BOST, BTPN, and BOBRN.
January 2007	Validity of 7500 km or 30-day close-circuit rakes extended to 35 days. 7500 km or 35-day close-circuit rakes extended to 10 close-circuit bases. Premium pattern permitted for all air-brake-stock except BRN, BTPGLN, BLC, BFN, and departmental stock.
April 2007	Premium pattern extended to total 90 examination points. Since then, further progress has been made.

Source: Taskforce on Train Examination Practices, Railway Board, Ministry of Railways, Government of India, 2008.

risk-averse organization. Second, early successes as in the case of an improvement in train examination practice, gave the staff confidence, allowing for a gradual scaling up. Third, mistakes, when they occurred, like in the case of subcontracting cleaning of a railway station which met with the disapproval of unions, had small fall-outs. Incrementalism allowed for hedging risks and taking immediate corrective measures, especially in politically contentious and institutionally critical issues like safety and labour. Further, risk-to-return trade-offs were considered. Initiatives with low risk and high returns were accorded precedence. For instance, over a four-year period only one-third of the total route kilometres were gradually approved for CC+8 heavy axle load freight trains; but these routes carry two-thirds of the traffic. Finally, incremental change needed small amounts of investment up front, making it affordable for a fiscally constrained Railways. In sum, a calibrated and deliberative approach allowed for consistent change ensuring a broad-based support because risk was minimized and rewards were quick and boosted staff confidence.

Chase

Chase, the mentality of a hungry tiger on the hunt, is an essential ingredient of the supply-side reforms. Analogous to the tiger, the senior management identifies new policy targets and chases and grabs them.[18] The chase has several elements. First among these was setting stretched targets. Once targets were set, the staff articulated the needed inputs to achieve the target. Here stretch and leverage are twins, and the latter is inspired by the former.

Second was to pursue the set of strategic inputs simultaneously. Increasing axle load required collaborative inputs from several departments. Here the principle driving chase was the spirit behind Robert Bruce and the spider, 'Try, try, try again, till you succeed.' This management by nagging toolkit consisted of the following. Memos with clear objectives were issued and directed to all relevant departmental heads, seeking a response in a pre-specified and realistic timeframe. Critical decision memos were stamped with 'top priority' in red ink, adding to the sense of urgency. Meanwhile, there was relentless follow-up over the phone, chasing the file from one desk to another—from the director to his joint secretary, further up to additional secretary, then finally to the board member. While the file moves five levels down in the Railway Board, it then finds its way up the same order, making it ten steps and a tedious process. Further, this was done across the concerned mechanical, electrical, civil, traffic, and finance departments. For each revision on axle load the files go through a similar process. Since there were over forty-five rate circulars issued on the matter, the files on increasing axle load moved through ten desks in each of the five departments forty-five times requiring inputs and approvals at 2250 desks. Moreover, in many cases inputs were sought from all the sixteen zonal railways and at times involved field visits. Just one input for the axle load required chasing over 2000 desks in four years, not to mention chasing all the zonal railways, which in turn chased hundreds of field units. Imagine the quantum of inputs required to reduce wagon turnaround time where over twenty inputs from several departments were simultaneously solicited. On several occasions there were differences between departments and this resulted in a deadlock. In response, the reformers organized meetings to facilitate exchange of views, meanwhile nagging with follow-up memos and supplementary information. Then, at a more personal level, the viewpoints of all factions were heard, files were chased on the phone, attempts were made to solve issues informally over tea, else at the level of the Board, and finally, if nothing worked, feedback and decisions were sought to meet the annual budget deadline. The complexity associated with running faster, longer, and heavier trains with hundreds of cross-functional and spatial inputs across the bureaucratic web was enormous. This was one of the primary constraints that had inhibited action in the past.

Third was timing, because in this context when to act is of essence. The most critical period is the annual Railway Budget that is treated in the Railways as a rigid deadline. Besides, there are two or three supplementary budget deadlines, as well as general manager's conferences and monthly board meetings. Each such deadline was deployed to focus the attention of senior management in order to resolve differences and make critical decisions. Further, senior officers are most receptive to new ideas and other inputs prior to their promotions and during the early days of their appointment. Put simply, high priority events, entirely based on the institutional culture and psyche of the organization, were leveraged to meet deadlines for deliverables. Thus, choosing when to raise the next generation of policy reforms was as important as identifying and prioritizing initiatives.

Fourth, change was induced by demand. By implementing decisions and following up with requisite inputs, a sense of urgency was created. For example, first train length was extended, then platforms were lengthened. The interim imbalance created by the extra length of a train that did not have access to a platform encouraged divisional managers to act swiftly to build the additional length of platforms. Similarly, axle load was increased before the weaker segments of tracks were replaced. In the interim, the trains were operated with speed restrictions on the weaker track segment. Not only did these demand-responsive approaches create a sense of urgency and induce the system to play catch-up, they also allowed for incremental earnings to precede incremental investment.

Finally, in contrast with the tedious process of decision making, when it came to implementation it was surprisingly swift. For certain decisions that translate directly into policy directives not much follow-up was required. Examples of such directives are the rate circular on tariff changes, change in axle load, or on revised train examination practices, that by default, became effective from the date specified in the notification—as soon as the decision was communicated to the zonal railways, it resulted in instantaneous implementation of the new practices.

Other decisions that required actual construction like overcoming terminal constraints and removing network bottlenecks were done in a timely manner, partially because funds were allocated, decision-making authority devolved, and demand induced. But more importantly speedy implementation resulted from the Railways's super-strong execution capabilities that rest on the shoulders of its technological prowess and follow-up mechanisms. Further, to leverage this sophisticated monitoring and evaluation system, the reformers integrated critical budgetary missions and reform targets into the existing system. In this context, Carlos Ghosn's insight regarding Nissan is equally applicable to Indian Railways. In the context of reforming Nissan, Ghosn observes that

as long as management gives clear directions that everyone understands, as long as you've got a clear, thoroughly explained strategy, you don't need to worry too much about how well and how fast it's carried out. Don't get me wrong, you still have to expend a lot of time and energy, but it's remarkable how execution falls into place [Ghosn and Ries 2005: 210].

PARSING SUPPLY—LONG TERM AND THE INTERIM

The essence of the supply strategy's temporal sequencing is captured in the Railway budget speech for the fiscal year 2007–08 (Budget Speech 26 February 2007).

The growing demand for transportation can only be met through a harmonious blend of short-term and long-term policies. Our short-term policy of investing in low-cost high-return projects has been successful in eliminating network bottlenecks and in ensuring effective utilization of rolling stock. Alongside a twin mid-term and long-term investment strategy will be adopted to enhance productivity through modernization and technological upgradation on the one hand and enhancement of capacity of the network and rolling stock on the other [p.12].

As seen in the preceding pages, the emphasis of this strategy is to augment capacity through a combination of expanding the railway network and increasing productivity by deploying technology. On the one hand, with the improved financial condition of the Railways, capital investment has increased from Rs 11,000 crore in 2004 to Rs 28,000 crore in 2008, implying a compounded annual growth rate of 30 per cent. In particular, the annual production of the rolling stock has been ramped up from 6300 in 2004 to 14,000 wagons a year in 2008. Equally, diesel as well as electric locomotive production has more than doubled from 116 to 222 and 86 to 200 a year respectively.

More than the mere increase in allocations and quantities is the improvement in quality. The new rolling stock is technologically superior and has been redesigned to enhance capacity of each new wagon, locomotive, and coach. The production of old low horsepower diesel and electric locomotives is being phased out over a five-year period ending in 2011—this long-drawn phasing is not due to liquidity constraints, but the need to test and develop a reliable and cost-effective supply chain of vendors for the new technology. Meanwhile, annual production of the new-generation diesel locomotives has been ramped up from 24 to 59, and electric engines from 16 to 55. Similarly, coaches have been redesigned to accommodate additional seats and better amenities for passengers; meanwhile the existing stock is being retrofitted to incorporate these improvements, particularly in the air-conditioned classes. The improved capacity of redesigned wagons is captured in the 2008–09 Railway Budget speech (Budget Speech, 26 February 2008).

With the objective of increasing the carrying capacity, from 2008–09 onwards manufacture of 20.3 ton axle load BCN and BOXN wagons will be stopped and only 22.9 ton axle load stainless steel wagons will be manufactured. The newly designed stainless steel BCN wagon has a lower tare weight. Due to the shorter length of these wagons, instead of 40 wagons, the BCN wagon train will now accommodate 58 wagons, like BOXN wagons trains. Thus, the payload of the BCN trains will increase by 78% from 2,300 tons to 4,100 tons. Similarly, the payload of open wagon trains will increase by 22% to 4,100 tons. We have achieved this by reducing the tare weight and increasing the width and height of the wagon [p. 18].

See Table 4.3 for a comparison between the old and new wagons and their payload capacities.[19]

Further details of redesigned wagons are provided in Appendix 5. Likewise, in the passenger segment the capacity of air-conditioned three-tier coach has been increased from 64 to 72, air-conditioned-chair-car from 66 to 102, air-conditioned first-class from 18 to 22, air-conditioned two-tier from 46 to 52, and the second-class sleeper coach from 72 to 84. Likewise there is an effort to enhance throughput by route-wise planning and development of the track network which is articulated in the Railway Budget 2008–09 (Budget Speech, 26 February 2008).

Sir, more than 75% of Railways's goods traffic moves on about 20,000 kilometer of the Railways's high density network, (namely) coal and iron ore routes, and port connectivity railway lines. Many of these routes are fully saturated and capacity utilization is in excess of 100%. …An investment of about rupees 75,000 crore (US$ 17.4 billion) will be made over the next seven years to augment line capacity on these routes. Route-wise works will be undertaken in a phased manner including 124 works of doubling, third and fourth-lines, bypasses, flyover, crossing stations, intermediate block

Table 4.3
Carrying capacity of old and new covered wagons (BCN)

	Old wagon	New wagon	Old rake	New rake
Tare weight per wagon	24.55 tons	20.8 (↓) tons		
Payload per wagon	58.6 tons	70.6 (↑) tons	2360 tons	4095 tons
Volumetric capacity per wagon	103.4	92.5 (↓) m³	4550 m³	5365 m³

signaling, automatic signaling works, and yard remodeling.…104 throughput enhancement works in progress would be completed over the next two years. This entire network will be provided with IBS (intermediate block signaling) by March 2009 [pp.12–13].

Finally, through dedicated freight corridors, multi-modal logistics parks, world class railway stations, and five rolling-stock factories, the long-term strategy is to anticipate and provide for future growth needs. The process for enhancement of throughput has been simplified—both approval and execution of these projects will require less time than other projects.

Through this supply-side management, Indian Railways, has grown consistently at 8 per cent annually. This growth is less due to an increase in the rolling stock or railway network, and more due to productivity gains from higher efficiency and effective utilization. Meanwhile, most of this period, 2004–08, has been characterized by macroeconomic stability with inflation hovering between 4 and 6 per cent and relatively low interest rates.[20] At real (constant) prices the Railways's unit cost has been declining since Independence in 1947. But in a historic first, even at nominal (current) prices the railways's unit cost declined due to productivity growth rates outpacing the rate of inflation. Therefore the freight unit cost declined by 12 per cent from 61 paise a ton in 2001 to 54 paise at current prices in 2008. The unit revenue from the freight business segment was 74 paise in 2001 and the profit margins were around 21 per cent. With a 7 paise decline in unit costs, the profit margin has nearly doubled from 21 per cent in 2001 to 37 per cent in 2008. Likewise, in the same period, the unit cost per passenger kilometre remained almost constant, increasing marginally from 38 to 39 paise. This miraculous improvement results from a supply strategy to play on volumes for reducing unit costs. As will be shown in the following chapter, this provided substantial room for demand-responsive pricing to increase market share and expanding profit margins. The discussion of the supply side preceded that of demand because several analysts have argued that a large proportion of the freight traffic has been lost not due to lack of demand, but due to capacity constraints. In this regard, the RITES report (1997: 2.24) argues that *matching capacity to requirements is critical*, 'In fact, wagons and locomotives are in short supply, most of the major routes are working to near saturation level of capacity and demand for rail movement is ahead of supply (p. 2.24).' However, once the supply-side management had increased the availability of wagons and coaches and reduced other supply constraints, a dynamic and differential pricing policy, and a market-driven and customer-centric response was required. Details of this demand-side strategy are the concerns of the next chapter.

5 Service with a Smile

The essence of the demand-side strategy is best captured in Mahatma Gandhi's observation on his visit to the Indian Merchant Chamber, Mumbai, sometime in the early part of the twentieth century.

A customer is the most important visitor on our premises. He is not dependent on us. We are dependent on him. He is not an interruption of our work. He is the purpose of it. He is not an outsider to our business. He is part of it. We are not doing him a favour by serving him. He is doing us a favour by giving us the opportunity to do so.

Like many things in the Mahatma's life, this too is hard to emulate, but the Railways has made a sincere attempt to reinvent itself as a customer-centric organization. This chapter provides insight into Indian Railways's struggle to make the shift from a statist and monopolistic organization to a demand-responsive—dynamic, differential, and market-driven—customer-focused organization.

LOOKING IN THE WRONG PLACE

In order to respond to customers' needs, the Railways began simplification and rationalization of its pricing policy. In steel, where it had been losing market share, freight charges were reduced by about 22 per cent from class 230 to 180. Moreover other incentives were introduced. Incremental freight traffic earnings, in comparison with the previous year, received a 15 per cent discount and authority to approve these discounts was devolved to the field units. Customers were offered loyalty discounts. Loyalty towards the railways was measured by the rail coefficient—namely railways's share of the total cargo of a given commodity with a particular customer. Additionally, major customers were also rewarded with quantity-based discounts. On the pricing front, all the tricks of the trade had been exhausted. Despite these efforts, the rail coefficient for steel kept declining from 67 per cent in 1991 to 35 per cent in 2005. What was counter-intuitive was that this decline occurred despite far lower rail fares as compared to trucks. With the objective of initiating a dialogue, seeking customer feedback, and becoming demand responsive, the Minister constituted a committee of the major freight customers under his chairmanship. In one meeting of this committee, the Minister expressed his frustration: *'Hum kiraya aur kitna ghatayen? Free kar den kya?'* (How much more should we reduce the fare? Should we make it free of charge?)

The CEO of a mega steel corporation retorted: 'You are looking in the wrong place.' He expalined that, the Railways provided a station-to-station service for transportation of steel and the incidental costs associated with rail transport were very high due to multiple handling, truck transport costs at both ends, ware housing, increased inventory, and the like. And these costs outweighed the savings accruing from cheaper rail charges because rail freight was a fraction of the total door-to-door logistics cost to his steel company. Therefore, for customers like him to opt for Railways's services would require the Railways to offer small batch consignments.

The Railway Board leapt up in defence, 'The Indian Railways cannot return to being a transporter of piecemeal traffic, carrying 10 tons of cargo per customer like the truckers do.' The representatives of the steel industry were quick to reply, 'No, we just need half train loads instead of full train loads (that is 2000 tons)'. The minister intervened, 'We grant you that, now will you shift your freight to the Railways?' 'Not really,' came the reply, 'we also need to unload at multiple locations on the way.' 'Granted,' responded the Minister. And the CEOs of the steel and cement companies asked in, 'How much will you charge for these services?' The Minister replied, 'Gratis, this is my gift to you.' The rest is history. Once the Railways started accepting half train loads and providing the option to unload in a combination of stations en route,[1] it succeeded in arresting and then reversing a sixty-year old trend of a declining rail coefficient for steel. Between 2005 and 2008, the rail coefficient for steel traffic increased from 35 to 45 per cent and for cement from 41 to 45 per cent. This reversal was due to a combination of market-driven policies tailored to the customers' requirements.

Quintessentially, through such client engagements the Railways had learned the basic principle of the market: market share is to be fought for and won in the market place. For winning this battle consistently, the customer should be offered superior and compelling value on a continuing basis. The ultimate measure of value for the Railways's services was customer satisfaction. To create this value, the Railways has transformed itself from an introspective organization where the emphasis was on process and procedure, to one that is externally oriented with emphasis on the market and the customer. But succeeding once does not mean that the customer can be taken for granted. The Railways learned this lesson the hard way. In fiscal year 2006–07, it was overconfident after gaining market share in steel and cement. In this exuberance it introduced a 5 per cent peak-season surcharge for the new services. In response, the customers for steel and cement shifted, and the Railways recorded negative growth in the freight for these commodities[2] in April 2007.

However, the Railway Board was resistant to any hasty revision: 'How can we change policies announced in the budget without waiting for its approval?' 'There has to be some stability, uniformity, and consistency in our pricing policies.' While their concerns were in place, and the Railways cannot be as volatile as the market, a degree of dynamism was essential.

Soon after the budget was approved by Parliament on 9 May 2007, the Minister invited the customers for a meeting at the Railway Museum, where the customers complained about the surcharge imposed on the mini-rakes and 'two-point' rakes as well as the incremental freight discount policy, 'With respect to incremental freight, how can we provide a quantum increase year-on-year? Why don't you asses the incremental freight discount from the base year of 2006?'

In response,[3] the Minister announced the revocation of the surcharge and revised the incremental freight discount policy with a fixed base year of 2006. But, in the interim, the damage was done. The Railways lost traffic in May, and then again in June, leading to poor freight loading in the first quarter. This was followed by the lean season. There was a lag due to existing freight agreements between customers and alternate service providers. But with the introduction of a lean-season discount and revision in the incentives, the customers began trickling back.

It would be misleading to draw the conclusion that all customer demands were met. Accommodations were made only where demand was relatively elastic and the threat of losing the customer to alternate modes was real. Thus, the railway reformers were selective in their response. For instance, in the customer meet discussed earlier, the corporate representatives complained about two things. First, about fare-hikes, namely 50 per cent increase in freight rates of iron ore as well as other minerals, and 33 per cent increase in foodgrains and fertilizer freight charges. Second, about termination of the minimum-weight condition. They now had to pay for load that they did not transport. While wagons could carry only 60 tons of urea, the customers were being charged the full carrying capacity of the wagon, which is 64 tons. But these concerns were not addressed because despite hefty increases in freight charges and the problem of dead weight in urea, these customers had not opted out. As seen in Chapter 3, the Railways is a door-to-door transporter in most of these market segments—like iron ore and coke—with negligible incidental costs. And rail freight is over 50 per cent less than comparable road freight. Likewise, customers demanded discounts on total load, that too year round, as opposed to discounts being limited to incremental freight in the lean season. These demands were also not met. Similarly, foodgrains and fertilizer corporations demanded mini-rakes and two-point unloading facilities gratis, but these demands were only partially addressed by offering these facilities at an additional surcharge of 5 to 10 per cent. Thus, the reformers conscientiously leveraged the Railways's relative competitive strength in these door-to-door freight segments. Nor was there any political backlash in the case of foodgrains and fertilizers. As seen earlier, this was a case of perceived political sensitivity rather than a real concern because the cost of freight was borne by the public exchequer and not the end consumer.

In essence, the demand strategy had four critical elements: it was differential, dynamic, market driven, and customer centric. The critical instrument was to respond with a combination of price and non-price initiatives where the Railways faced a real competitive threat and thus offer superior value at competitive prices. As a first step the tariff structure was rationalized.

RATIONALIZATION OF TARIFFS

Until 2005 the Railways had an encyclopedic and multi-volume tariff schedule spread over 500 pages containing over 4000 entries for different commodities. These entries included traditional Indian sweets like rasgoolla, balusahi, jalebi, laddoo; types of hair like camel and human; musical instruments and players like the tape recorder, gramophone records, gramophone needle cakes, pianos, and the like. This tariff schedule not only specified the names of commodities but further sub-classified them. For instance, it included 261 sub-categories of cotton. And the rate classification for these 261 sub-categories ranged from class 130 to 240 with thirteen different classes in all.[4] To further complicate matters, freight was charged on the basis of the minimum-weight condition—referred to as 'W' commodities in the schedule—or based on actual carrying capacity of the wagon—referred to as 'CC' commodities. Various attributes of commodities—processed or raw, hard or soft, powdered or granular, compact or loose—were relevant for determining the loadability of each commodity in different types of wagons. For instance, cooking coal is low in density and therefore was classified as 'W', that is to be charged based on the minimum-weight condition, while ordinary coal is dense and therefore was classified as 'CC'. Further, for coal there were twenty-four entries with seven types of differing 'W' weight conditions. The rate classification ranged between 130 and 165 with four different classes in all. In sum, the permutations and combinations of these categories created an unfathomable rate matrix. These classifications were a cause of confusion leading to mistakes in billing to customers as well as collusion to misclassify freight to benefit from lower rates for similar commodity types. Neither railway staff nor customers could distinguish between the various categories and identify the precise sub-category of the commodity prior to each loading. This ambiguity created a cesspool of corruption.

All these were inheritances of the past when the Railways was in the business of transporting piecemeal traffic—referred to as smalls and wagon loads.[5] In 1981 the Railways decided to end piecemeal traffic and accept mostly train loads.[6] As a result, thousands of these categories became redundant but the tariff schedule was not revised. For the first time since 1958, a comprehensive revision of tariff was undertaken in fiscal year 2004–05. An ABC analysis[7] revealed that eight major commodities accounted for more than 85 per cent of freight traffic. Further, seventy-one commodities accounted for over 97 per cent of the freight traffic. Hence all obsolete entries of the bygone era of small and piecemeal booking were deleted.

Second, all commodities were clubbed under twenty-four generic group heads, for example all kinds of alloys and metals or all types of chemicals and fertilizers were clubbed under one head.[8] While earlier different alloys and metals could have varying freight rates, the revised tariff schedule provides uniform rates for all commodities within a major group head, unless specified otherwise.

Third, in the revised tariff schedule the minimum-weight condition was scrapped. The Railways as a transporter decided to charge according to the load that it can carry—the carrying capacity of a wagon—as opposed to the load that is actually being carried. This method of charging for carrying capacity has been the norm in the freight industry. For instance, cabs charge for the distance and duration of the trip irrespective of the number of passengers travelling. This rationalization of the tariff structure, in one stroke eliminated the cumbersome red tape, systemic corruption, and suboptimal use of wagons. The hegemony of the goods clerk over the customer due to the 'plethora of imponderables' that allowed enormous discretionary powers was eliminated with the revised tariff schedule. Increased transparency, and its simple to use and enforce format, have reduced harassment of the customer on the one hand and tariff evasion on the other.

RATE RATIONALIZATION

While the goods tariff specifies the applicable rate class for the commodity to be transported, the rate table provides the applicable charge for the rate class, identified in the goods tariff, for a specific distance. In the past, the freight rate table[9] had a class range of 40 to 300, and the ratio between the minimum and the maximum rate class was 1:8. Post rationalization and simplification, the ratio between the minimum and maximum freight rates has been narrowed down to two, with the minimum class set at 100 (that is the break-even price) and the maximum at 200 (which is two times the break-even price). Further, the number of classes have been reduced from 27 to 11 and now have a uniform increment in multiples of tens. In the pre-rationalization rate table the class interval between 40 and 190 was in increment of fives and beyond 190 was in increments of tens and the taper varied by classes. Additionally, the taper for deriving the telescopic rates—decreasing block tariff—is now uniform across all classes, whereas earlier it varied among classes.

In the passenger business, while the Mohan Committee (2001b: 72) had recommended decreasing the ratio between the highest and lowest passenger classes from fourteen times down to ten largely by increasing the fares of the non-air-conditioned classes, the rail reformers achieved the same objective by decreasing the fares of the highest classes. This was both commercially prudent and socially optimal because the Railways gained market share in its high-end, high-margin segment while the poor customers were not burdened. Such, a win–win solution faced no political resistance.

Finally, essential services like defence and postal tariffs that had traditionally been subsidized by the Railways were revised upwards so as to cover costs of operations and a reasonable profit. Likewise, the fares of special trains operating for marriages and political rallies have been hiked to reflect the cost of service.

DIFFERENTIAL PRICING POLICY

In the socialistic era, all pricing was affordability based—low-value commodities and poor passengers were charged less while higher-value commodities and travel segments were charged higher fares. Despite liberalization of the Indian economy in 1991, this pricing policy remained unchanged because fares were considered to be politically sensitive. Since ministers resisted fare hikes in politically sensitive segments—suburban, ordinary passenger, and second class mail and express trains—the brunt of rising costs, more often than not, was borne by high-value finished goods and air-conditioned travel classes. This eroded the Railways's competitiveness in these segments. In particular, steel and cement customers migrated to trucks and air-conditioned class travellers to budget airlines. Yet fares for these segments kept increasing and the Railways was consistently pricing itself out of the market.

There was a clean break from an affordability-based pricing policy[11] and a monopolistic approach to the market. The policy of announcing across-the-board price hikes to make up for the budget deficits, independent of the Railways's competitiveness, was scrapped. The current pricing strategy is differential and customer centric. The pricing policy based on the socialistic principle of affordability has been creatively modified. While affordability-based pricing continues to guide the politically sensitive second class travel segment, the pricing policy for the entire freight, parcel, and air-conditioned business segments is now completely market driven. Currently, in these segments, fares are increased or decreased depending on the Railways's competitive edge. For instance, in order to regain competitiveness in the passenger business, air-conditioned first class and two-tier fares have been reduced by 28 and 20 per cent respectively. As seen in Chapter 3 the Railways lacks a competitive edge in station-to-station freight segments while it has a formidable edge in the door-to-door segment. But in the affordability-based pricing regime, station-to-station services—largely high-value finished goods like steel and cement—were charged more while the door-to-door service, low-value commodities like iron ore and other minerals, were charged less. This has now been revised; freight charges for low-value door-to-door commodities like iron ore have been increased by 50 per cent. On the other hand, with a view to impart a competitive edge, station-to-station freight transportation rates have been reduced or kept constant. For instance, over the last few years freight rates for petroleum products and steel have been reduced by 33 and 22 per cent respectively. Several non-price incentives are also being offered to improve the quality of service—such as mini-rake loading, multiple-location unloading. In sum, the effort has been to strengthen the total value offered to the customer. As a result, the Railways has been gaining market share in freight, in both door-to-door as well as station-to-station services.

Yet another exemplar of differential pricing policy is the *empty flow direction scheme*. In the past pricing policy did not make a distinction between loaded and empty-flow directions, and freight charges were the same for both. As a large proportion of the Railways's trains return empty, the 'empty-flow direction freight discount scheme' was outlined in the Railway Budget (Budget Speech, 26 February 2007) to capture some traffic in the empty returning freight trains.

Sir, the truck rate for Delhi to Guwahati is considerably higher than the rate for the return trip whereas the Railways charge the same rate in both directions. It is seen that 40 out of 100 freight trains return empty. The additional expenditure in loading freight in the empty flow direction trains is quite low. Hence, I announce a heavy discount on incremental freight in the empty flow direction. For distances beyond 700 kilometers, the discount will be 30 per cent during non-peak season and 20 per cent in the peak season. The scheme will be applicable for all items loaded in covered wagons. In the case of open wagons, the discount will be applicable for all commodities except coal, coke, and iron-ore for export. In peak season, this discount will be applicable for open wagons for distances over 1000 kilometers only [Budget Speech, 26 February 2007: 327).

In the subsequent budgets empty-flow rebates were increased to 30 per cent and made applicable round the year. Authority was devolved to the General Managers of the zonal railways to increase the discount up to 50 per cent. Incremental freight traffic requirements were also relaxed for loading at good sheds—but not from private sidings. Field units could offer discounts on total freight traffic, as opposed to incremental loading, in the empty-flow direction for loadings at goods sheds. Likewise, for loading of foodgrains, fertilizer, and cement in open wagons, usually transported in closed wagons, double discounts were provided in the empty-flow direction. The customers were offered the usual empty-flow discount as well as a compensatory discount to make up for the decreased loading capacity of open wagons and the floor[11] to this combined discount was set at class 70.

Finally, the new *Tatkal Seva* is a fee-based service where passengers can buy tickets last minute. In several trains, as soon as advance reservation services become available, that is ninety days prior to the travel date, all tickets get sold. In the past, passengers had to approach touts for buying tickets at a premium. Now the Tatkal service, has been extended to 30 per cent of the total seats in a train and is offered over a period of five days prior to the travel date. The service has differential prices for higher and lower travel classes, and the prices are dynamic between lean and peak seasons and popular and less-popular trains. Not only has the Tatkal Seva stymied touts while serving the needs of last-minute travellers but it also rakes in Rs 300 crore a year for the Railways, another classic case of a win–win outcome.

Dynamic Pricing

After the initial success of the Railways's reforms with rationalization of the tariff structure, the reformers turned towards market-oriented freight rates and passenger fares. Dynamic aspects of the pricing policy are illustrated through variability introduced by the reformers between peak and lean seasons. This is summarized in the budget for fiscal year 2005–06.

Railways's passenger fares and freight rates remain unvarying for all seasons and for all routes, whereas tariffs in the airline and road sectors vary depending upon the demand and the season. In order to be able to effectively face the challenges posed by stiff competition, in the current year we had started a discount scheme for non-peak season and empty flow direction for freight rate, which has been successful. As an extension of this policy, I propose to introduce a Dynamic Pricing Policy for freight as well as passenger, for peak and non-peak seasons, premium and non-premium services, and for busy and non-busy routes. As per this policy the rates for non-peak season, non-premium service and empty flow directions will be less than the general rates and the rates for peak season and premium services could be higher than normal. For the freight the non-peak season would be 1st July to 31st October. For the passenger segment this period would be 15th January to 15th April and 15th July to 15th September [Budget Speech, 24 Febraury 2006: 31].

In the lean season, demand declines and truckers lower their freight charges. The Railways also responded with decrease in prices, specially in the station-to-station segments[12] during the lean season. The Railway Budget outlines the modalities of 'the non-peak season incremental freight discount scheme'.

The demand for freight transportation dips from 1st July to 31st October on account of monsoon. Hence, during this period, under non-peak season incremental freight discount scheme, freight rebate of 15 per cent will be offered for incremental freight revenues of over rupees five crore in a month and ten per cent if the incremental earning is less than rupees five crore. This rebate will be applicable for all commodities except coal, minerals, and items with classification below 120 [Budget Speech, 24 February 2005: 32].

Conversely, in the peak season, when demand exceeds supply, 5 to 7 per cent busy season surcharges were levied. Likewise, in the passenger segment, pricing for the air-conditioned segment became dynamic—less during the lean season and in unpopular trains. Since there was no political space to increase passenger fares in the peak season or for popular trains, even in the air-conditioned classes, dynamic pricing was achieved by lowering fares in the lean season and for unpopular trains.

In the past, in classes where occupancy rates were low, seats were left vacant and the Railways lost revenue. In an effort to have customer-centric and market-driven passenger services, an automatic upgrading scheme has been introduced and now travellers from a lower class are upgraded to the next class in case of vacant seats—an empty seat in air-conditioned first class is filled by a air-conditioned two-tier passenger and so on. Railway staff do not have discretionary powers to choose whom to upgrade, four hours before the departure of the train, upgrading is done by software that randomly selects travellers. As a result the Railways not only has higher occupancy per train but also gets revenue for an otherwise lost seat. Customers too are delighted at the possibility of an upgrade. Here too, there are some exceptions. For instance, on Shatabdi trains most sales are last minute and over the counter prior to the departure of the train. Thus these trains do not have an upgrading scheme. In essence, the reformers focused on macro responses to demand like seasonal variations or to-and-fro variations. Some micro management like time-of-day convenience pricing and auctioning of vacant seats and berths in passenger trains as well as empty returning freight trains is also in the process of being implemented through a commercial portal. Auctioning will be done at scheduled prices or lower because the objective is to increase utilization and occupancy rates and not profiteer from scarcity.

Price Discovery

Each passenger train, with the exception of short-distance computer services, has two brake vans—one in front and one at the end—with a total parcel-luggage carrying capacity of 16 tons per train. Further, the Railways has about 800 parcel vans that are attached to passenger trains as per demand. For this parcel service, the Railways's rates are set in three brackets—highest for Rajdhani and Shatabdi, followed by mail and express, and then ordinary passenger trains; these are denoted by R, P, and S classes in that order.

However, in the parcel business the Railways is a station-to-station transporter. Therefore, in the short lead the Railways is not competitive. There is hardly any demand for parcel booking in the frequently stopping, short-distance, ordinary passenger trains and over 50 per cent of mail and express trains that travel a distance less than 750 km. Overall, the Railways's competitive edge increases with distance and reliability of the service—punctuality, time of service, and the like. Further, brake and parcel vans going from production to consumption centres are in demand, but on the return trip these vans are empty. As a result, only 20 per cent of the total parcel capacity was being utilized.

Despite two years of trial and error, the reformers were unsuccessful in improving parcel business. Through a process of learning by doing, the reformers discovered that in the parcel segment, speed and reliability, not price, are of essence. Parcel charges per kilogram of cargo between Delhi and Mumbai vary between Rs 2 and 12 depending on how consistent and quick the service is. As soon as the Railways aligned its price to the industry norm of speed and reliability, business improved. A critical element of the parcel business revival strategy was initiating a wholesale leasing of the brake and parcel vans through open competitive bidding. However, the parcel operations of the Railways had deep-rooted vested interests that were resistant to wholesale leasing, citing concerns of redundancy of porters and the parcel clerks. Thus, initially outsourcing was introduced in one brake-van per train. Gradually, these price discovery mechanisms were extended to include a second brake-van. The rest of the parcel segment continued the past practice of routine piece-by-piece booking of parcels.

Once the freight forwarders' willingness to pay was assessed through competitive bidding, all attributes affecting demand—

time, speed, reliability, and directional flow—were built into the price. Several long-distance trains, were leased at rates substantially higher than the scheduled parcel rates, but piecemeal booking continued at much lower scheduled rates for the remaining brake-vans. In response, the pricing policy was revised such that if brake-vans of a passenger train are leased out at more than the scheduled rates, piecemeal booking by the Railways is done in the immediately higher class, provided leasing operators honour the lease agreement for at least a period of one year. For example, if a train is leased at higher rates than P—the intermediate rate—after a year, piecemeal booking will be done at R—the highest rate band. With this decision, earnings from piecemeal booking increased significantly; in some cases earnings doubled and the hold of vested interests weakened.

On the other hand, there was no response on scheduled parcel rates for several trains. In such cases, field units were authorized to progressively reduce the reserve price from 100 to 75, 50, and even 25 per cent of scheduled parcel rates, subject to a minimum of the previous year's earnings. This process led to an increase in capacity utilization of the brake and parcel vans. As a result, the Railways's parcel and luggage earnings more than doubled between 2004 and 2008 from Rs 476 crore to 1008 crore.

ALLIANCES FOR VALUE CREATION

Like in the case of the parcel business, other alliances for value creation were also sought where the sum of the parts was greater than the whole. In this regard there were three underlying principles that guided the partnerships. First, co-optation of competition was sought in sectors where the Railways was a minority or declining transporter. Second, alliances were looked for to align long-term interests of the existing customers with those of the Railways. Third, alliances were also forged to improve the quality of service and enhance the overall value proposition for the clients. For instance, to co-opt competition, fifteen container-train licences have been issued to firms from the logistics industry—shipping, road transport, as well as warehousing. These players have added over seventy trains to the existing 140 container trains operated by the Container Corporation of India (CONCOR), a public sector undertaking of Indian Railways. While CONCOR built this fleet of trains over two decades, the private players have added half as strong a fleet in two years. The container business is now growing at twice the earlier rate at 24 per cent annually. These new players have been incentivized to focus on adding new customers rather than diverting the existing freight customers of the railways, thus creating an expanding pie scenario. The concessions are for twenty-two years and thus foster long-term partnerships.

To foster additional long-term alliances the Railways initiated investment schemes for wagons and rail sidings and incentivized engine-on-load. Under these schemes the initial investments are made by customers—to build sidings or produce wagons—and a portion of the cost is reimbursed by the Railways through discounts in total freight billing over a ten- to fifteen-year period. Under the wagon investment scheme, clients were invited to invest in their own wagons. Depending on the type of wagons, discounts in freight charges are granted—these range between 10 and 15 per cent for a period of ten to fifteen years. This is to reimburse the customer investment in wagons along with interest. As a result, not only is the customer committed to using the railway freight service for the long term, but the Railways also gets additional investments in wagons (Budget Speech, 26 February 2008).

Further, with the objective of tying in the customers for the long term and providing railway connectivity within the customers' premises, the Railways revised its policy for construction of sidings. Unlike the past, where all capital costs were borne by the customer, now half the cost is borne by the Railways and reimbursed to the customers via a discount in the freight over a period of ten years or more. Further, salaries of railway staff posted at the sidings were earlier borne by the customers. Now, except for one commercial staff per shift, all other costs are borne by the Railways (Budget Speech, 26 February 2005). As a result, the Railways provides customers a door-to-station service instead of a station-to-station service.

Under the engine-on-load schemes, for quicker release of wagons the engine stands by during loading and unloading operations. In the past, the engine would place the rake and return after the loading was over. While loading and unloading took up to a day or more, the engine also lost time commuting back and forth. Under this new scheme, the permissible time for free-of-charge loading and unloading has been reduced to four hours for open wagons and six hours for covered wagons, as opposed to nine hours in the past. Reduction in time spent on loading and unloading requires investment in modernization and mechanization of handling equipment and infrastructure, which is being reimbursed to the customer. This scheme is outlined in the Railway Budget for fiscal year 2006–07:

Customers who fulfill the conditions laid down in the scheme and invest in their terminals so as to bring down the loading and unloading time, and complete loading or unloading in lesser time, will qualify for five per cent rebate in the first year. Over the next ten years the rebate will be given at a diminishing rate and would be one per cent from the fifth year onwards [Budget Speech, 24 Febrary 2006: 33].

INNOVATION

To complement the tariff rationalization, rejigging of the product mix, and efforts to improve the quality of service, the Railways introduced several innovative products and services. Mini-rakes, two-point unloading, Tatkal Seva, and automatic upgrading of passengers are some examples of recent innovations. However, by far the most popular and novel product is the Garib Rath, the poor peoples' chariot. This is the Nano of Indian Railways, except that it hit the market a few years earlier. Quite like the Nano car launched by Tata Motors, the Garib Rath has four integrated attributes, namely affordability, scale, aspiration, and efficiency. It provides air-conditioned travel at affordable prices—about half the passenger fares of an ordinary Delhi–Mumbai three-tier air-conditioned coach. Figure 5.1 captures the essence of a strategy to decrease the unit costs by increasing the number of coaches per train as well as efficiently using space within each coach to accommodate more passengers in the Garib Rath as opposed to the normal Rajdhani train. As against seventeen coaches in a normal train, the Garib Rath has twenty-four coaches. Further, unlike the air-conditioned three-tier coach in a normal Rajdhani train that accommodates sixty-four passengers, the Garib Rath accommodates seventy-five passengers. Likewise, the chair car of a Rajdhani has space for seventy passengers, whereas its equivalent in the Garib Rath has space for 102 passengers.

Therefore, while a standard train carries 816 passengers, the Garib Rath accommodates more than twice the number of passengers with a capacity of 1920. Further, as most costs are fixed, and thus insensitive to the number of passengers, the unit cost per traveller decreases substantially from 80 paise in a Rajdhani Express train to 32 paise in the Garib Rath as summarized in Table 5.1. Through a combination of the above attributes and removal of all paraphernalia—non-paying coaches like the pantry cars—Garib Rath tickets are priced at about half of the Rajdhani air-conditioned three-tier fare. Further, there are no concessional tickets, no discounts, not even for rail pass holders. Thus it meets the aspirations of the common people of India who could not, until

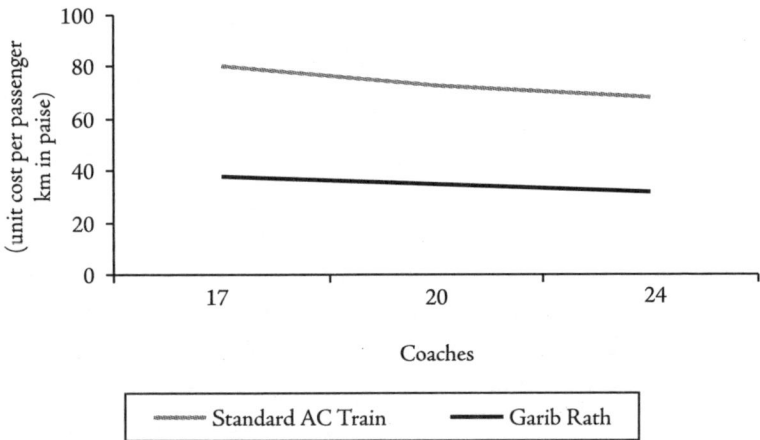

FIGURE 5.1: Effect of train length and coach layout on unit cost

Table 5.1
Economics of a Garib Rath train as compared to
a normal Rajdhani

	Normal Rajdhani train	Garib Rath
Seating capacity three-tier	64	75
Chair car	70	102
17 coach train	816	1233
24 coach train	1302	1920
Cost per passenger	80 paise	32 paise
kilometre	(17 coaches long)	(24 coaches long)

Source: Statistics and Economics Directorate, Ministry of Railways.

now, afford air-conditioned travel. These mutually reinforcing attributes not only make the Garib Rath affordable to the poor but also hold the potential to revolutionize the generally loss-making segments of the passenger business. Thus, this is an exemplar of the win–win outcome that was discussed in the Chapter 2 on the political economy of reforms.

Profit Margins and Product Mix

Profit margins in the freight business have improved from 20 per cent in 2001 to 80 per cent in 2008. A simplistic assumption may be that this implies the misuse of monopoly power to price gouge. However, this is not the case. Freight rates of coal and cement, that account for nearly half of freight earnings, have increased marginally, while those of steel and petroleum products, accounting for another 15 per cent of freight earnings, have been reduced. Thus 65 per cent of commodity freight prices, barring busy-season and development surcharge, have been reduced or remain constant. The radical improvement in profit margins is attributed to the triple combination of reduction in unit costs, selective fare hikes, and transformation of low-margin business segments into super-high-margin ones.

Declining unit costs is the first factor contributing to this improvement. In 2001, the Railways's freight unit cost was 61 paise per net-ton-kilometre with a unit realization of 74 paise.

Thus, even during the worst of times, the Railways enjoyed a profit margin[13] of 21 per cent in the freight business. During the seven-year period between 2001 and 2008, even after absorbing inflationary pressures, the unit cost has fallen by over 11 per cent from 61 to 54 paise at current prices. Even at the past level of unit revenue, profit margins would have expanded from 21 to 37 per cent solely by virtue of declining unit costs. Conversely, if unit costs had continued to increase as per the past trend of 8 per cent (the compounded annual growth rate between 1991 and 2001), they would have increased to 103 paise in 2008. The Railways would have to increase freight rates by 40 per cent to break even and by 70 per cent to retain the 20 per cent profit margin. This illustrates the importance of declining unit costs in fuelling an exponential growth in profit margins. The decline in unit costs is a function of macroeconomic stability characterized by low inflation—hovering between 4 and 6 per cent as opposed to 10 and 12 per cent over the 1990s—as well as double the growth rate of 8 per cent in freight loading.

Second, between 2001 and 2008 unit revenue increased from 74 to 93 paise. This is partly due to selective increases in freight rates of door-to-door commodities—namely iron ore and minerals where the railway has a formidable competitive edge by about 50 per cent, and of low-rated commodities like foodgrains and fertilizers, which have been underpriced in the past by 33 per cent. Moreover, levying the development and busy-season surcharges at the rates of 2 and 5–7 per cent respectively, led to further increase in revenues.

The third contributory factor is transforming low-value, low-margin iron ore for export business segment, to a high-value high-margin one. Success in business is about spotting an opportunity, seizing, it, and cashing in on it. Booming prices of iron ore from US$ 20 per ton in 2004 to over US$ 100 in 2008 in global markets offered the Railways such an opportunity. Iron ore is mined in the central plateau regions of India where trucks are unviable—the road conditions are poor, slopes are steep, and the commodity is bulky in nature—and therefore price elasticity of demand is low.

Taking this into consideration, freight rates for iron ore for export have been quadrupled.[14] Even at four times higher prices, there is no dearth of demand—over 10,000 indents are pending in South Eastern Railways—and transport volumes have increased from 36 to 53 million tons. These fare hikes have no societal implications either, because the commodity is for export and prices are pegged based on global demand, nor are the customers opting out because of the plum profit margins they enjoy resulting from the sudden increase in global iron ore prices. This is a classic example of revamping a low-margin and low-value business into a super-high-margin and high-value one. As a result, freight earnings from iron ore for export are estimated to increase from Rs 900 crore in 2004 to 9000 crore in 2009 and its contribution to the overall freight earnings is expected to increase from 3 to 13 per cent in this period (see Table 5.2).

In fiscal year 2008–09, freight rates of all commodities other than iron ore for export have not been increased. Yet, growth in freight earnings substantially exceeds growth in volumes in the first five months of 2009. This is because of a 115 per cent growth in earnings from the freight of this one commodity alone which has increased the overall growth in earnings by 8 per cent—from 13 to 21 per cent.

TABLE 5.2
Contribution of iron ore for export to total freight traffic

	2004	2008	2009*
Iron ore for export's share of total freight earnings (%)	3	9	13
Growth in earnings of iron ore for export (%)	52	63	115
Revenue from iron ore for export (Rs)	900 crore	4400 crore	9000 crore
Growth in freight earnings without iron ore for export (%)	3	10	13
Growth in freight earnings with iron ore for export (%)	4	14	21

Source: Statistics and Economics Directorate, Ministry of Railways.
Note: *Based on data up to July.

Further, passenger business offers yet another illustration of leveraging the varying growth rates in different business sub-segments to accelerate growth of earnings at a pace that exceeds the growth in volumes. Over the last four years, the passenger volume recorded a compounded annual growth rate (CAGR) of more than 6 per cent while earnings had a much higher compounded annual growth rate of 12 per cent. This was achieved despite a reduction in passenger fares for most classes. A simplistic explanation offered by sceptics is that there has been a clandestine increase in fares. But the counter-intuitive differential growth rates of revenue and volumes is explained by the differential growth in high-margin and high-value services versus low-margin and low-value services (see Table 5.3).

As seen in Chapter 3, suburban services account for 57 per cent of the total number of passengers but contribute only 8 per cent of the total passenger earnings. On the other hand, air-conditioned plus mail and express non-air-conditioned sleeper segments account for 4 per cent of the total number of passengers but account for 46 per cent of total revenue. This paradox of differential growth in volumes and earnings is explained by a 10 per cent growth in the volumes of high-value segments while there has been a 5 per cent growth in the low-value segments. For the former, namely air-conditioned and mail and express sleeper, each

Table 5.3
Changing product mix and revenue growth

	Distribution of Travellers (%)	CAGR (2004–8) of Number of Travellers (%)	Revenue per Passenger (Rs)	Share of Total Revenue (%)
Air-conditioned service	1	10.5	638	20
Mail and Express Sleeper service	3	9.6	215	26
Suburban service	57	5.4	4	8
Total	100	6.4	28	100

Source: Statistics and Economics Directorate, Ministry of Railways.
Note: All figures are for fiscal year 2006–07, unless mentioned otherwise.

passenger on average pays Rs 638 and Rs 215 respectively while for the latter, namely the suburban segment, each passenger on average pays Rs 4. In essence, price is only one variable among many that affects growth in earnings, product mix is another critical variable.

IMPROVING QUALITY OF SERVICE

To improve quality of customer service, several complementary efforts—namely improving operational efficiency through investments in new technological and human resources, investments in amenities, systemic changes through deployment of information technology, and strategic partnerships—were made.

Reliability, punctuality, safety, productivity, operational efficiency, and profitability are organically interdependent. These are complements not substitutes. Improvement in reliability and safety not only reduces the damage to assets but limits the disruption of Railways's operations, thus increasing punctuality, which in turn improves productivity and profitability. For instance, the recent reduction in turnaround time, a key gain in productivity, is a result of significant improvements in reliability, punctuality, and safety. With this faster turnaround of wagons, not only are customer demands met quicker—as seen in shorter waiting-lists of indents[15]—but also the Railways's assets are being better utilized and thus profitability has increased. Therefore, quality of service and productivity are inherently interdependent and the following five vectors were central to the strategy. To address client grievances, moreover, regular redressal meetings are being conducted at various levels of the system.

First, the Railways is making massive investments to upgrade its technology and modernize the rolling stock, signalling and telecommunications, tracks, and other assets for improving reliability, safety, and operating efficiency. For example, the Railways has more than tripled its allocations for the depreciation reserve fund from Rs 2300 crore to Rs 7000 crore between 2001 and 2008. These investments have been complemented by investments in human resource development and information technology.

As a result, the number of accidents have declined to less than a half—from 320 in 2004 to 194 in 2008—and asset productivity and operating efficiency have made significant gains. The freight customer benefits from reduction in inventory during transit, timely delivery, improved reliability of the service, and reduced damage and pilferage. This, makes it a win–win outcome with customer satisfaction rising and improved utilization of Railways's assets. However, there is much more that can be done and to this end the Railways is investing in initiatives that will bear fruit in the near future.

Second, the Railways has made significant improvements in passenger conveniences through investments in amenities. The height of most platforms has been modified to match the height of trains, covered shelters have been provided in hundreds of small stations, and there are no constraints on funding improvements in passenger amenities—like, better illumination and drinking water facilities—and the general ambience as well. Likewise, major goods sheds are being renovated and access roads and other facilities are being improved.

Third, the customer interface is being improved through systemic changes. For instance, through a revision of the tariff schedule and scrapping of the minimum-weight condition, the procedure as well as documentation has been simplified. Likewise, information technology has been deployed. In the past customers had to deposit either cash or demand drafts. This led to a lot of last minute stress at the customer's end. Now sitting in the 'comfort of their office', freight customers can avail the e-payment facility. And the Railways is in the process of installing its online freight operating information system for all major customers so that they can track their trains in real time.

In the passenger segment, to improve the customer interface—across the value chain from ticketing to travel—internet ticketing services have been developed by the Railways's own Indian Railways Catering and Tourism Corporation. Twenty per cent of all reserved ticketing is done through these portals and online bookings have doubled since the last year, albeit from a small base. Further, e-tickets are available at over 40,000 outlets across the country from petrol pumps to ATMs, bank counters, and several chain stores and small shops. This has not only reduced the queues at ticket counters but also improved the customer's experience in buying tickets.

In the past, customers were either unable to reach train enquiry, or when their call was answered, their experience was unsatisfactory. Now, customers dial 139 for the *Rail Sampark* service, a year-old nationwide railway enquiry system. The service is provided in eleven languages by a joint venture between a business process outsourcing firm and a telecom firm. From anywhere in India telephone enquiries can be made at the cost of a local call. Among other things, this service provider responds to queries pertaining to arrival and departure of trains, reservation status, fares, and passenger name record (PNR). The service is provided through four call centres and has recorded an exponential growth with over half a million calls being answered each day. Further, the world over, call centres are cost centres, but in this innovative and perhaps unique arrangement, the call centre operator is not only paying for the capital and operational expenses but also pays a small annual fee to the Railways. The operator recovers the capital and return on investment by sharing revenue with the telecom operator as well as earnings from the provision of several value-added services like SMS alerts, and hotel and cab reservations.

Fourth, in an effort to further deepen improvements in the quality of service, the Railways sought external partnerships from private as well as public enterprises. Introduction of catering kiosks and food plazas managed by national and multinational corporations including Haldiram's and McDonald's to improve the quality of food served at stations is an example. Further, to improve transit accommodations at major railway stations, the Railways's Yatri Niwas, previously loss-making outfits have been contracted out to The Taj Group's Ginger hotel chain and other such players. Train toilets have an improved ambience because private firms like Airtel have been given advertisement rights along with the responsibility of improving ambience and upkeep of the coaches. The cleaning of toilets on trains has been contracted to Eureka Forbes. This was done without displacing the existing cleaning personnal who now play the role of cleaning inspectors and thus are happy with the introduction of the new service. On stations, the Railways has been leveraging eyeballs, footfalls, and stomach bowls to enhance non-fare passenger incomes as well as improve the quality of the transit experience. Through these alliances, the Railways's sundry earning have more than doubled from Rs 1000 crore to Rs 2600 crore in the last four years. This is another example of a win–win outcome, because on one hand the Railways has enhanced its profitability and on the other customers receive an improved service.

In conclusion, the entire inclusive reform strategy and efforts to transform the Railways was tested in the market place as customer satisfaction was the final measure of the total value of the Railways's service. As long as the Railways continues to innovate and add value that results in customer satisfaction, this transformation will sustain. However, there are several threats to sustainability as well as lessons to be learned for replication that are examined in the next chapter.

6 Outcomes, Sustainability, and Replication

In this chapter some aspects of the Railways's achievements are summarized. These outcomes are not just improvements in the Railways's finances but tangible gains in capital and labour productivity, increase in market share and profit margins, better quality of service, as well as enhanced stature of the Minister. Further, a critical assessment is made of the perceived and real concerns for sustainnig this transformation. While most critics have raised concerns in regard to passenger safety, short-term gains at the cost of long-term ones, and cashing in on a booming economy, the critical threats to sustainability seem to lie elsewhere. Finally, some potentially transferable attributes of the Railways's success are summarized and for purposes of illustration the electricity sector is considered in brief.

OUTCOMES

The scale of the financial transformation of Indian Railways is best captured in the following. In 2008, the Railways had a cash surplus before dividend of Rs 25,006 crore (US$ 6 billion), operating ratio of 75.9 per cent, and a fund balance (or bank balance) of Rs 22,279 crore (US$ 5.2 billion). Moreover, the ratio of net revenue to capital-at-charge (return on net worth) had improved from 2.5 per cent in 2001 to 20.7 per cent in 2008, making it a relatively high return on equity for a capital-intensive infrastructure industry like the Railways. The Railways has improved its debt service–cash coverage ratio over three times from 1.74 in 2001 to 6.53 in 2008. In the financial sector, after the American sub-prime crisis, there is a global credit squeeze and lenders are risk averse, specially when it comes to emerging markets like India. Despite these developments, in November 2008, based on a much improved balance sheet, the Indian Railway Finance Corporation placed dollar-denominated bonds with the Bank of Tokyo, Mitsubishi for US$ 100 million at 4.01 per cent (LIBOR [2.56] + 1.45 per cent), an interest rate which is lower than the cheapest rate offered to many of the Fortune 500 firms around the world. Table 6.1 contrasts the Railways's performance on several financial indicators summarizing the scale of change (see Appendices for detailed financial results).

Over the 1990s the Railways's expenses grew 1 per cent faster than its earnings, leading it towards bankruptcy. However, between 2001 and 2008, the Railways has become super-solvent by inverting this relationship—now earnings grow 4 per cent faster than expenses (see Table 6.2). This results from a combination

TABLE 6.1
Financial indicators

(in Rs crore)

	2001	2008	Change
Cash surplus before dividend	4790	25,006	5-fold increase
Investible surplus	4204	19,972	Over 4-fold increase
Capital expenditure	9395	28,680	3-fold increase
Fund balance (bank balance)	359	22,279	62-fold increase
Operating ratio	98.3%	75.9%	22 % improvement
Ratio of net revenue to capital-at-charge and investment from capital fund (return on net worth)	2.5%	20.7%	18 % improvement
Debt service–cash coverage ratio	1.74	6.53	Over 3-fold increase

Source: Finance (Budget) Directorate, Ministry of Railways.

of factors. Earnings have grown on account of growth in freight volumes, selective fare hikes in previously underpriced freight business segments, as well as change in product mix in favour of high-value high-margin segments. Meanwhile, working expenses grew at a lower rate primarily on account of low inflation and the Railways's relatively inelastic costs structure with respect to volumes transported. While such gains in freight have been the backbone of this transformation, the outcome for the Railways as a whole is similar. Table 6.2 compares the growth in earnings and expenses over the 1990s that led to the financial crisis in 2001 with the transformation thereafter.

Moreover, the growth of earnings has doubled in the last four years—from 7 per cent in 2001–04 to 14 per cent in 2005–08—and, on average, the gap between the growth rates of earnings and expenses has also doubled—from 2.5 per cent between 2001 and 2004 to 5 per cent in the following four years (see Figure 6.1).

In the same vein, the gains in the recent four years exceed the gains in the four years that preceded them. Between 2001 and 2004, traffic earnings recorded a 7 per cent growth, while working

TABLE 6.2
Compounded annual growth rate of expenses and earnings

	1991 Rs crore	2001 Rs crore	2008 Rs crore	1991–2001 CAGR %	2001–08 CAGR %
Total working expenses	11,154	34,667	54,462	12.01	6.67
Gross traffic receipts	12,096	34,880	71,720	11.17	10.85
Passenger earnings	3148	10,515	19,844	12.82	9.50
Goods earnings	8408	23,305	47,434	10.73	10.69
Other coaching earnings	336	764	1800	8.56	13.02
Sundry earnings	242	703	2565	11.25	20.31

Source: Statistics and Economics Directorate, Ministry of Railways.

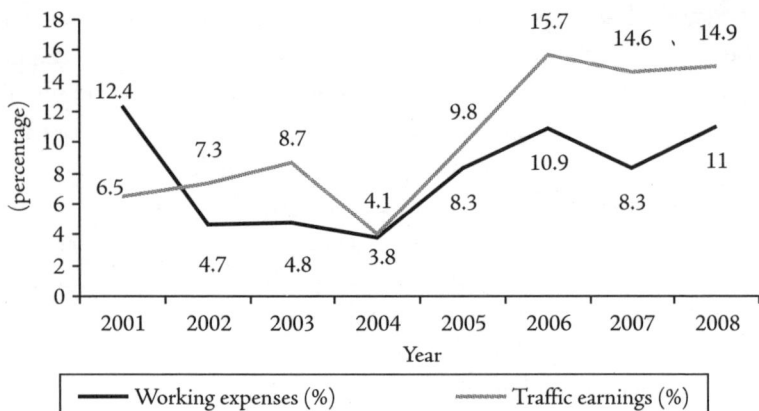

FIGURE 6.1: Growth in traffic earnings versus working expenses

expenses grew at 4 per cent, and as a result the investible surplus grew by about 40 per cent from Rs 4200 to 5800 crore (US$ 1.4 billion). However, between 2005 and 2008 these gains multiplied three times to Rs 20,000 crore (US$ 4.7 billion). In sum, a few years after the predicted financial crisis, Indian Railways became one of India's most profitable enterprises with US$ 4.7 billion in profits—namely investible surplus (see Figure 6.2). While the Railways's profitability is not directly comparable with that of privately owned corporations, the scale of its achievement is significant.

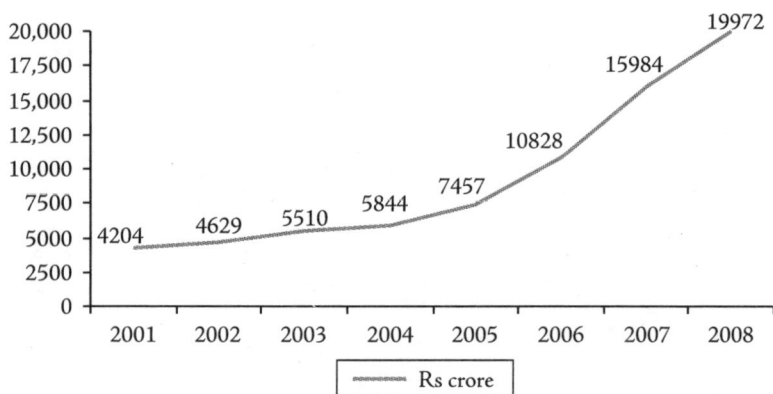

FIGURE 6.2: Growth in investible surplus

There were modest gains in the investible surplus between 2001 and 2004 as freight unit revenue fell from 74 paise in 2001 to 72 paise in 2004 (see Figure 6.3). This was largely on account of reduction in freight rates for petroleum products and steel to curtail its declining market share in these commodities. However, the reformers discovered that this was insufficient to regain market share because these commodities, as seen in earlier chapters, are more sensitive to non-price factors such as quality of service—like options to transport less than train loads through mini rakes as well as unloading en route at more than one location. Further, the Railways was hesitant to introduce selective fare hikes in underpriced commodities like iron ore, foodgrains, and fertilizers because of the misplaced conception that these were politically sensitive. Hence, despite a reduction in unit costs from 61 paise in 2001 to 57 paise in 2004, the Railways's freight operating margins made moderate gains. Likewise, gains in passenger earnings in 2001–04 were low, compared to 2005–08 because passenger unit costs increased from 38 paise to 41 paise and business strategies such as rejigging the product mix in favour of high-value high-margin air-conditioned and long-distance passenger segments were yet to be explored (see Figure 6.4).

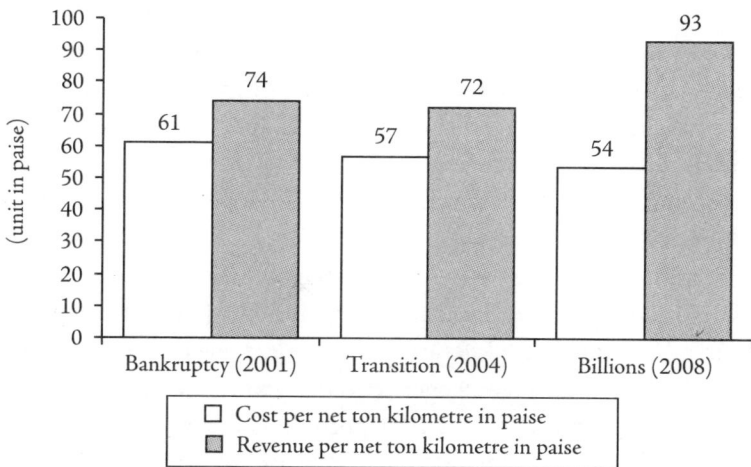

FIGURE 6.3: Freight unit revenue and cost

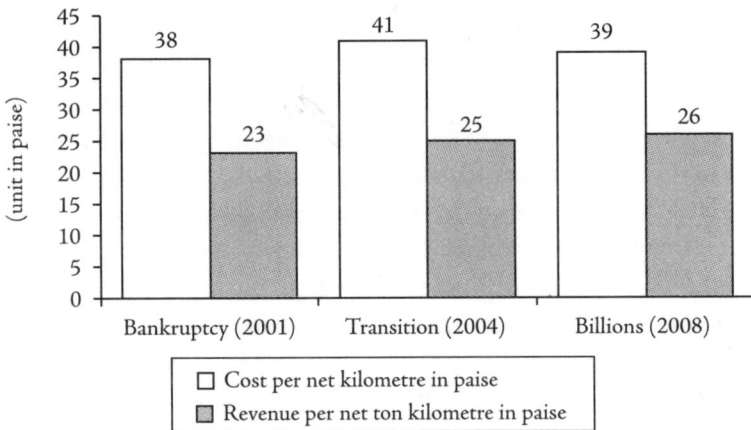

FIGURE 6.4: Passenger unit revenue and cost

Investible surplus during 2005–08 has grown exponentially because of growth in freight volumes, declining unit costs (from 57 to 54 paise), and increasing unit revenue due to selective fare hikes (from 72 to 93 paise, see Figure 6.3). Despite reduction in passenger fares of most travel classes, losses in the passenger business have been curtailed because of two factors. First, marginal fall in unit costs (41 to 39 paise) and, second, increase in unit revenue due to a change in product mix in favour of high-value and high-margin air-conditioned, and long-distance travel segments (25 to 26 paise, see Figure 6.4). Further, the growth rate of 'other coaching' as well as 'sundry earnings' has increased from around 10 per cent in 1991–2001 to about 15 per cent in 2001–08. This results from enhancing non-passenger fare income through leveraging eyeballs and footfalls of travellers and by reduction in unutilized parcel capacity.

Similarly, the growth rate of freight and passenger traffic volumes nearly doubled during the period 2001–08 as compared to decade of the 1990s. Likewise, over the last seven years following 2001, the number of passenger trips has grown two times than in the 1990s. And the freight transported, as measured in net ton kilometres, grew at 2.6 times the rate of the preceding period (Table 6.3).

TABLE 6.3
Compounded annual growth rate of output

	1991 Millions	2001 Millions	2008 Millions	1991–2001 CAGR %	2001–08 CAGR %
Passenger trips	3858	4833	6558	2.28	4.46
Passenger kilometres	295,644	457,022	767,519	4.45	7.69
Freight loaded in tons	318.4	473.5	794.21	4.05	7.67
Freight transported in net ton kilometres	235,785	312,371	511,801	2.85	7.31

This substantial growth in traffic volumes has been achieved with the same network, rolling stock, and employees—implying a significant gain in capital and labour productivity as summarized in Table 6.4.

What is surprising is that to achieve this financial transformation the Railways did not resort to textbook solutions of shrinking the *denominator*—that is cost cutting through retrenchment and

TABLE 6.4
Gains in productivity

	1991	2001	2008	1991–2001 CAGR %	2001–08 CAGR %
Wagon utilization (NTKM/wagon/day)	1407	2042	3566	3.79	8.29
Track utilization (NTKM/route kilometres in million)	3.78	5.01	8.09	2.86	7.09
Track utilization (PKM/route kilometres in million)	4.74	7.25	10.99	4.34	6.12
Labour productivity (NTKM/employees in millions)	0.15	0.22	0.37	3.90	7.71
Labour productivity (PKM/employees in millions)	0.19	0.32	0.55	5.35	8.04

privatization. Instead, the focus was on growing the *numerator*—namely increasing revenue by expanding volumes and earnings. As a result, while the proportion of staff costs to expenses remains stable, their share as a percentage of revenue has declined sharply. Staff costs were 54 per cent of operating expenses in 2001 and 50 per cent in 2008. However, the share of staff costs to gross traffic receipts declined from 51 per cent in 2001 to about 36 per cent in 2008 (see Table 6.5). This is primarily on account of rapidly rising volumes and earnings, and associated reduction in unit costs.

Further, the Railways's market share in the freight of finished goods like steel and cement has grown—the Railways measured this through the rail coefficient, a measure that monitors its freight market share in each of the eight major commodities. This is an important outcome because it had been losing market share in these segments for a decade and a half—market share for steel and cement in 1991 was 67 and 59 per cent respectively, while in 2004 it had declined to 36 and 40 per cent. In a historic reversal, the Railways increased its market share in the freight of both these commodities (see Figure 6.5). While the steel and cement industries grew by 8–10 per cent in 2008, rail freight grew by 25 per cent, resulting in an increase in the market share by 2 to 5 per cent.

Another critical outcome is improvement in customer orientation as measured by reductions in price and improvements in safety and quality of service. Along with the reduction of passenger fares in several travel classes, freight fares for several station-to-station commodities like petroleum products were lower in 2008 when

TABLE 6.5
Break-up of ordinary working expenses (Gross)

All amounts in crore rupees	2001	2008
Gross traffic receipts	34,880	71,720
(staff cost as percentage of GTR)	(51.21%)	(36.33%)
Ordinary working expenses	33,161	49,924
(staff cost as percentage of OWE)	(53.86%)	(50.20%)
Staff cost	17,861	26,059

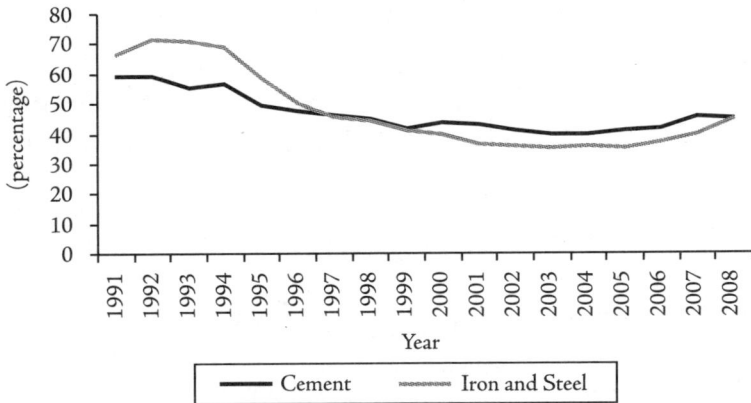

FIGURE 6.5: Railways's market share in steel and cement

compared to 2001 levels. Additionally, safety has improved as captured in the halving of the total number of accidents—from 473 in 2001 to 194 in 2008. Moreover, this has occurred while the total number of train kilometres had increased. Thus accidents per million train kilometres have declined by two-thirds—from 0.65 in 2001 to 0.22 in 2008. Quality of service for passengers and freight customers has improved as a result of systemic changes such as e-ticketing, complementary upgrading of passengers if seats are vacant in upper class coaches, contracting out cleaning services on trains, improving the ambience of railway stations through private participation, e-billing for freight customers, and a host of other service improvement initiatives discussed in earlier chapters. Finally, the transit time for travellers and freight customers has declined and field units of the Railways have greater autonomy to resolve customer grievances.

What did the Minister and railway staff gain through these positive outcomes? Based on a performance ranking of all the federal ministers by a leading news magazine, at rank 33 Lalu was at the bottom of the list, in 2005 (*India Today* 2006). In 2006 the same magazine ranked him as the second best Minister of the coalition government. During his tenure as Minister of Railways, Lalu has been able to transform his image as well. Earlier he was associated with failure of governance in Bihar and now the media refers to him as a 'Professor Lalu' after he gave a lecture to business school students at IIM Ahemdabad as well as students visiting from Wharton and Harvard. In sum, the Minister has enhanced his stature. Likewise, breaking away from the fear of retrenchment and uncertainty associated with the financial crisis, the Railways's transformation has given a boost to the morale of the railway staff. In 2007, in a ranking of public sector organizations in India by NDTV, a leading national news service provider, the Railways stood second, just below the Indian Army, a significant achievement for an organization that was close to bankruptcy in 2001. In essence, the outcome of this Railways's transformation have been multifaceted—improvement in its functioning by gains in productivity, generating a substantial cash surplus, benefiting the customers, and strengthening the political leadership as well as boosting the staff morale, without burdening the common people of India. However, this is not to say that the Railways has surpassed its own potential or that the expectations of policymakers and customers have been met. There is a lot more scope for improvement, making the sustainability of this transformation a central concern is going forward.

Sustainability

Policymakers and academics have raised five major concerns regarding this transformation. First, the Railways's improved performance is a result of creative accounting. Second, financial gains are at the cost of safety. Third, growth in revenues is a result of a rapidly growing economy. Fourth, transformation of the Railways is a result of plucking low hanging fruits, as in easy fixes, and thus short lived, particularly when it faces severe capacity constraints— super-saturated high-density networks, terminal constraints, energy shortage, and lack of wagons and coaches. Fifth, the fate of the transformation, after the present political leadership departs, is bleak.

These concerns are substantially misplaced. Concerns around creative accounting mainly stem from two factors: first, the pace and scale of the financial gains the Railways has made—from bankruptcy to a annual cash surplus of over US$ 5 billion in a few years; second, changes in the Railways's accounting practices introduced between 2005 and 2008. Regarding the first concern, the Railways's accounts and financial statements are verifiable like those of any other public enterprise and are open to public scrutiny. The Railways's cash surplus is deposited with the Reserve Bank of India and is open to verification. The second concern is about modification in accounting practices with respect to the allocation of leasing charges, interest on fund balances, and reimbursement of losses by the public exchequer stemming from operation and maintenance of strategic railway lines. Earlier, all leasing charges were recorded in accounts as revenue expenditure. But in the revised procedure, the interest component of the leasing charges is recorded in operating expenses (Rs 2153 crore) while the principal component (Rs 1677 crore) is recorded as capital expenditure. This accounting change amounts to a difference of Rs 1677 crore or US$ 390 million in 2008. Likewise, earlier reimbursement of losses on account of strategic lines (Rs 637 crore or US$ 148 million) was recorded in accounts as a reduction in the dividend liability. Now it is included in sundry earnings. In the same vein, interest on fund balances (Rs 1175 crore or US$ 273 million), which was not recorded as earnings earlier, is now included in total earnings for calculation of cash surplus before dividend. These changes are in sync with corporate accounting practices and are based on the recommendations made by external experts on railway accounting reforms. As a result of these accounting changes, the Railways's cash surplus records an incremental Rs 3489 crore (US$ 811 million) in 2008. However, to eliminate ambiguity, all past and present cash and investible surplus data used in this book are computed based on the current accounting practice and are thus comparable. Thus modifications in accounting practice do not account for the financial transformation.

Regarding the safety–profitability trade-offs, in contrast to popular perception, safety, productivity, and profitability have a complex interdependence and are thus complements not substitutes. Thus, it should not be a surprise that the number of train accidents has declined from 473 in 2001 to 194 in 2008, while the railway's profitability has soared. Further, during the run up to the financial crisis in 2001 the Railways had a cash crunch and was unable to afford replacement of overaged assets. In contrast, for the fiscal year 2008–09, assets, internal generation, and non-budgetary resources are contributing 78 per cent of an annual plan outlay of Rs 37,500 crore (US$ 8.7 billion) that not only covers cost of replacing aging assets but also investment needs to meet future demand. Moreover, all major operational changes having safety implications in the railway are strictly scrutinized by the Railways's Research, Designs, and Standards Organization (RDSO) as well as an independent Commissioner of Railway Safety, administered by the Ministry of Civil Aviation, Government of India.

Third, while the Railways's freight business benefits from surging demand in a rapidly growing economy, it is not sufficient to yield higher earnings and profits. Between 1987 and 2001 the Indian economy grew at over 6 per cent, but rail freight grew at about 2 per cent. In contrast, between 2005 and 2008 the Indian economy grew at 8 per cent, and the Railways, through a series of interventions discussed earlier, cashed in on this surging demand by expanding its freight business at 9 per cent per annum. To recapitulate, cashing in on this demand required structural improvements in the functioning of the Railways to alleviate supply-side constraints so that the Railways could respond. Further, between 2001 and 2008, both local and global macroeconomic stability, characterized by low inflation and interest rates, greatly enhanced the Railways's ability to reduce its unit costs at current prices. A combination of high inflation and high interest rates can disrupt this virtuous cycle of growing volumes and falling unit costs. Moreover, commodity cycle upturns have benefited the Railways. While spiking petroleum prices imposed an additional burden of about a billion dollars on the Railways, surging demand for commodities like steel, cement, and iron ore helped expand its freight business segment manifold. As discussed earlier, high international prices of iron ore contributed a few billion dollars of additional earnings to the Railways through increase in freight charges on exports. However, the volatility of the commodity market and its impact on the Railways's freight business is best captured in the wake of the recent global financial crisis and credit crunch. As the commodity cycle unwinds, the Railways's freight volumes decline and freight earnings from iron ore for export alone wipe out a billion dollars in expected earnings.

Fourth, the quick fix argument, or gains from low hanging fruits, does not stand up to scrutiny either. Back-of-the-envelope calculation to benchmark the Indian Railways with Chinese and American railroads provide a glimpse into how underutilized the Railways's existing assets are. An equally large state-owned Chinese Railways has the same number of passenger kilometres, but transports four times more freight than its Indian counterpart. Likewise, the freight only, class one American railroads has one-tenth the number of employees as compared to Indian Railways while it transports three times the amount of freight. In essence, the scope to improve productivity in Indian Railways's is enormous.

Moreover, to meet future demand, targets are set for the Eleventh Five Year Plan (2008–12), namely a 50 per cent growth in passenger and freight volumes. This will require a jump in investments to extend the network and expand the rolling stock, adopt new technology, plus improve the quality of rolling stock, signalling, and other systems (see Table 6.6).

Furthermore, Table 6.7 provides a comparison of past as well as presently planned investments for capital expenditure based

TABLE 6.6
Comparison of traffic volumes in past and present Five Year Plans

	10th Plan 2003–07	11th Plan 2008–12 (projected)	Per cent increase
Freight traffic (million tons at origin)	728	1100	51
Net ton kilometres (billion)	481	702	46
Passenger trips (billion at origin)	6.3	8.4	33
Passenger kilometres (billion)	696	942	35

Source: Planning Commission, Government of India, 2008.

TABLE 6.7

Comparison of investment sources in Five Year Plans

(in Rs crore)

Source of investment funds	10th Plan 2003–07	11th Plan 2008–12 (projected)	Per cent increase
Gross budget support (per cent of total)	37,516 (45%)	63,635 (27%)	70
Internal generation	29,567 (35%)	90,000 (39%)	204
Extra-budgetary resources	16,981 (20%)	79,654 (34%)	369
Total	84,064 (100%)	233,289 (100%)	178

Source: Planning Commission, Government of India, 2008.

Note: in the Eleventh Plan additional investments through public–private partnerships are anticipated.

on the five year plans. There is a concerted effort to increase reliance on internal generation, market borrowings (extra-budgetary resources) as well as public–private partnerships (not listed in the table but expected to raise US$ 4 billion) as opposed to relying largely on budgetary support from the public exchequer.

In an effort to increase the effectiveness and efficiency of investments, the construction works as well as the acquisition of rolling stock have been conceptualized with a commercial orientation to meet the demand of the logistics market. The investment strategy has been temporally parsed. Short- and medium-term fixes have been discussed in the preceding chapters. For instance, the Railways strategically invested in short-gestation high-return projects like increasing axle load on one hand and lengthening platforms to accommodate longer trains on the other. Further, several operational and procedural constraints were alleviated by modernization of loading practices and conditions on goods sheds as well as fixing network bottlenecks. Likewise, in the medium-term, three- to five-year period, several transit junctions were decongested by constructing flyovers and bypasses, goods sheds were provided with additional loading and unloading lines, and gauge conversion on some congested routes to serve as alternate paths. Yet these measures are not sufficient to meet the anticipated demand for traffic. For the long haul, along with the introduction of capacity- and efficiency-enhancing technology, systems, and procedures, the Railways will be investing tens of billions of dollars over the next five years for enhancing capacity. The Railway Budget has allocated resources for eliminating network bottlenecks, developing multi-modal logistic parks, and improving railway stations. These investments are being financed through a combination of internal generation, borrowings from the market, and public–private partnerships including outsourcing though long-term contracts (see Table 6.7). All this investment is being strategically channelled to projects that have a commercial orientation. For instance, through route-wise planning the entire high-density network's capacity will be augmented on a priority basis over the next five years at a cost of Rs 75,000 crore (US$ 17.4 billion). Likewise, priority is being accorded to strengthening iron ore and coal routes so as to carry 25 ton axle load. Finally, to enhance capacity in the long term, dedicated freight corridors are being developed along the length and breadth of the country to match the golden quadrilateral and its diagonals. To meet a surge in demand, factories are being built to manufacture engines, wagons, coaches, and their parts. Simultaneously, through public-private partnerships, billions of dollars will be invested in information technology. The central objectives of the information technology projects are to better inform senior management as they plan for the long term, for effective and efficient utilization of human and physical capital, reducing operating cost, increasing passenger and freight earnings, enhancing customer satisfaction, and further building brand value. In essence, Indian Railways is investing through a multi-pronged strategy to sustain its growth trajectory.

Finally, sustaining the management impetus after Lalu and his team leave office is less of a concern as the policy reforms have been embedded in the institutionalized DNA by instating systemic and procedural reforms. These have become part of the organizational routine, manuals, and to some extent norms. This can largely be attributed to leading change while respecting and strengthening the organizational identity and morale of the employees. Through a consensus-based approach, the reforms have developed deep roots within the institution.

However, on the organizational front several challenges lie ahead. Chief among these are attracting and retaining talented staff, creating a culture of cross-functional collaboration, increasing accountability across the hierarchy of the organization, and spurring innovation. First, in a reversal of past trends when the Indian Railways was a sought-after employer for highly qualified young professionals including IIT and IIM graduates, with rapid expansion of the economy, the Railways is increasingly facing competition from private employers.

In an ever-evolving business environment, recent efforts by the Railways to train officers in institutions outside the Railways, in premier universities both in India and abroad are a first step towards strengthening the capacity of existing staff. But it is not sufficient to meet the Railways's future needs and such capacity-enhancing collaborations should be further developed. There have been efforts to decentralize decision making to railway zones and within zones to divisions so as to increase the level of accountability of field units but there is a lot more to be desired on this front as well. In this vein, other institutional changes, like reorganizing the Railway Board based on the Railways's business segments, as opposed to the present functional silos, hold the potential to break departmental divisions and bring a cohesive business orientation to the organization as a whole. Increasing accountability across the hierarchy of the organization as well as enhancing the quality of work for the staff are goals that need further effort. Finally, to sustain these financial gains the Railways needs to foster innovation in commercial, operational, and pricing policies. Further, there is a need to promote synergy across zones and system optimization through cross-functional cooperation and spatial coordination. To encourage creative and nimble responses to an ever-shifting market dynamic, an innovation promotion group has been constituted in the Railway Board but this is still in its infancy and needs strengthening.

On the business front, it is essential to continue the focus on market and business segmentation and respond to firms' supply chains as well as distribution networks and client needs through tailor-made strategies. For example, a steel plant on the one hand needs to transport raw materials from the mines to the factory and on the other to transport finished products to the consumers. These are two distinct transportation needs of firms and the Railways needs to align its freight business in order to directly respond to these diverse business needs within the supply chain of firms, tailored to specific commodities like steel, cement, and fertilizer. For instance, to achieve the mission 200 million tons for the cement industry, a goal set in the Railway Budget (2008), the management needs to focus on articulating specific responses to the logistical needs of the specific supply chains. This includes, but is not limited to, developing tailor-made wagons for fly-ash, clinker, and bulk cement, and expanding as well as strengthening the rail network and terminals in new cement-producing areas. This industry-specific response is increasingly becoming the modus operandi and needs further development. Moreover, there is a need to anticipate capacity expansion by firms especially as several commodity producers, like steel and cement plants, plan to double capacity over the next few years.

Production of cement is expected to increase from 170 MT (million tons) to 280 MT by the end of the 11th Plan. Railways receive more than 100 MT traffic every year from the cement industry and we are targeting a loading of 200 MT from cement industry in 2011–12. There are more than 10 big clusters of cement production in the country. Work is in progress on Nandyal-Yerraguntla, Jaggayyapet-Mallacheruvu and Vishnupuram–Janpahad new lines and these will be completed in a time bound manner. The work of Bhanupali–Bilaspur–Beri new line in Himachal Pradesh has been proposed in the budget. The work of Daund-Gulbarga doubling and electrification of Pune–Guntakal line is proposed to be taken up to meet the demand of cement manufacturers in Wadi cluster. Gauge conversion and extension of Bhuj-Nalia line will be taken up after obtaining necessary approvals. More than 50 big terminals are being upgraded to increase their capacity, prominent among which are Mumbai, Pune, Chandigarh, Ghaziabad etc [Budget Speech 2008: 47].

To respond to the rapidly expanding steel industry, the Railways in its most recent budget articulated its own strategy to deal with the anticipated demand.

Steel production is expected to increase from 55 MT to 110 MT by the end of the 11th Plan. The Railways receives 120 MT traffic from the steel industry every year and we have targeted traffic of 200 MT from the steel industry by 2011–12. Most of the new dedicated iron ore routes will be constructed or upgraded for 25-ton axle load and some routes will be made suitable for running 30-ton axle load trains [Budget Speech 2008: 46].

In essence, the Railways does not only need to improve services to the existing commodity freight market by expanding its market share. It also needs to pre-empt future customers, tying them in through contractual arrangements that create long-term relationships as has been initiated through schemes like the wagon investment scheme and engine-on-load scheme, as well as incentives for building new sidings and rewards for incremental freight. Dynamic, differential, and customer-oriented policies by their very nature need perpetual revisions and refinement to match the market dynamics. For example, the Railways recently has introduced distance-based, also known as lead-based, differential pricing for cement and iron ore. Such strategies need to be adopted for other business sub-segments as well. Distance-based pricing takes into account the Railways's relative competitiveness. For instance, while long lead traffic, greater than 400 km, of iron ore for export is presently offered a discount of up to 50 per cent because international commodity prices have plummeted, short lead traffic, less than 300 km, of cement is being offered discounts of up to 40 per cent because of stiff competition from truckers. Such commodity-specific pricing strategies need to be dynamic and differential and increasingly tailored to specific commodity and geographic regions so as to reflect market conditions—including demand for commodities and credible threats from alternate modes. Another instance of a critical business segment that requires tailor-made responses by the Railways is the freight to and from the ports that accounts for a quarter of the Railways's freight business. Thus, there is a need to invest in eliminating bottlenecks along the freight corridors that carry this freight so as to increase the productivity of these high-density networks . This has been planned for in the Railway Budget for 2009 under mission 300 million tons for port traffic.

Indian Railways receives about 25% of the total traffic from various ports. India's foreign trade is likely to increase from 650 MT to 1100 MT by 2011–12. Thus, Railways is giving top priority to port rail connectivity projects (Budget Speech, 2008, p. 43). … Under the present scheme, during the concession period of 30 years, the Special Purpose Vehicle is eligible for proportionate net income or a return of 14% on equity whichever is less. The cost of investment in the construction of the project is based on actual expenditure incurred. To prevent time and cost over runs and facilitate real price discovery, it has been decided that on a pilot basis implementation of some projects will be explored on BOT (build-operate-and-transfer) basis through open tenders. The beneficiaries of the new line will give traffic guarantees [Budget Speech 2008: 45].

In sum, the threats to sustainability stem from macroeconomic volatility more than any other factor because it is beyond the scope of the railway to hedge for inflation and interest rate fluctuations. On the other hand, it is relatively better geared to tackling issues such as attracting and retaining talented staff, investing to meet future needs, as well as innovating so as to compete effectively in the market for transportation services. However, for an organization as large and complex as Indian Railways it would to be too simplistic to assume that all issues concerning the future of the railways have been addressed here or indeed can be anticipated. Thus, like other large organizations, the Railways's viability will be a function of its ability to learn from its past as well as constantly innovate to meet new challenges posed by a rapidly changing logistics market place. For instance, meet the increasing needs of containerization and multi-modal logistics that neatly integrate into global supply chains, while at the same time, successfully meeting the needs of poor commuters.

Replication

This successful transformation of Indian Railways contains some transferable lessons that may be considered in the context of reforming other public utilities. The conventional prescriptions of corporatization, privatization, retrenchment, hike in user fees, and independent regulation have been effective in sectors where charging and recovering user fee is not politically contentious as in the case of ports, telecom, and aviation. A classic case in point is the gains from private competition in the telecom industry in India.[1] However, there is a need to rethink this textbook approach to reforms in sectors like energy, water supply, irrigation, urban bus transport, and railway where charging and recovery of user fees, such that operation and maintenance costs are recovered, are politically infeasible. For such sectors, some ingredients for reinventing a reform strategy are identified. There has been no 'silver bullet' for reforms because they are always contextual and, in order to be effective, require detailed analysis pertaining to the sectoral, temporal, and spatial contexts in addition to the usual economic, political, and social contexts. These are not meant to be prescriptions, instead they are offered as possible ingredients of a strategy to tackle efforts to improve public service delivery in other network infrastructure sectors and perhaps in some other public service delivery as well. In this regard, a few striking lessons are outlined.

First, *inclusive reforms require a productive politico–bureaucratic interface* where the bureaucracy respects the political mandate and in return the political leadership refrains from interfering with the routine functioning of the bureaucracy. It is not uncommon for the bureaucracy and politicians to hold each other in contempt. Mutual contempt, or at best scepticism, needs to be replaced by mutual trust and understanding. This can be achieved by delineating the political domain of crafting the democratic mandate as distinct from the bureaucratic realm. The bureaucrats accept the mandate and channel its functioning to achieve the organizational objectives—providing quality service while having a profit motive—without compromising the political mandate.

Second, *commercial objectives and social obligations can be reconciled.* Counter-intuitive as it may be, the experience of the Railways's transformation demonstrates this. The Railways achieved this by first dissecting business segments into political and apolitical components, followed by further disaggregating it into nano constituents so as to identify apolitical variables that can be manipulated to improve profitability without compromising the interests of the political constituencies—in the case of the Railways it was the poor consumers and railway employees. In another sector this may differ. For instance, in the electricity sector in India, on disaggregating the consumer segments, it is quite likely that in most states electricity for industries, affluent neighbourhoods, and shops can be market driven, while electricity for rural and urban poor will require social considerations. And the subsidy can be targeted to these poor consumers alone. While the tariff structure attempts to make these distinctions in practice, there is a lot of scope to seek nano constituents of the consumer base that can be priced dynamically and differentially without burdening the poor. To translate such insight into action requires working across departmental silos, introducing a commercial orientation in the organization, and breaking free of a monopoly mindset. For instance, efforts to build cross-subsidies within the tariff accompanied by increasingly unreliable quality and quantity of electricity supplied, has led to the migration of consumers that usually pay their bills to independent power providers or install self-generation through captive power plants, both big and small, for industry as well as large residential enclaves. This opting out of the system in favour of alternatives is quite like the railways customers who, instead of paying high freight charges or air-conditioned travel fares, chose alternatives that offered better value. This has not been due to political interference but due to a lack of commercial orientation and a monopoly mindset within the state electricity boards.

Further, on the investment front, wheeling power to far-flung villages is a small fraction of the total capital expenditure of the sector where political considerations dominate. The rest can and should be commercially managed. For instance, there is vast scope in the electricity sector for improving utilization of existing capacity in generation, transmission, and distribution through low-cost, short-gestation, rapid-payback, and high-return investments. Similarly, bulk metering of electricity along the primary, secondary, and tertiary distribution networks requires small investments but promises quick and high returns. Separating technical losses in transmission and distribution from theft and making strategic investments to reduce unaccounted for power to large consumers like urban households and industrial areas can yield results quite like increasing axle load and lengthening platforms as the Railways did. Further, the widespread obsession with construction, procurement, and expenditure needs to give way to effective and efficient utilization of existing assets for enhancing productivity. In essence, an in-depth business and political analysis is a prerequisite for crafting an effective strategy such that apolitical variables are identified and managed on market principles. This example is only to provoke policymakers to explore such issues in another sector and are in no way prescriptive. What such analysis reveals is that there is immense scope for expanding desirable win–win outcomes where social and commercial objectives are simultaneously met.

Third, *efficiency must be improved by optimizing an underutilized system*. Across sectors there is enormous room to fix loopholes and reduce revenue losses. Several components of electricity generation, transmission, most of the distribution, as well as billing and collection can be market driven yet publicly managed, assuming that there is a will to override interest groups. As mentioned earlier, state electricity boards in India can substantially reduce aggregate technical and commercial losses that stand at 35 per cent, which is three times the global industry average, by improving billing efficiency that stands at 70 per cent and collection efficiency that is 94 per cent.

Fourth, *there is need to think anew*. While this may sound rather obvious from an enterprise perspective, many public utilities are grounded by inertia. In a fast-changing external environment, it is critical for public utilities to question past assumptions about the nature of the business, its cost structures and pricing, and respond appropriately. It is reasonable to assume that earning profits is not one of the central objectives of a public utility. On the contrary, in the Indian context, there is a profit aversion. As the experience of the Railways's transformation demonstrates, public utilities have to be reoriented to pursue profits. This is because a viable and financially sound utility is a prerequisite to meeting social obligations and objectives. This requires not only a new mindset but business savvy as well—the ability to spot, seize, and cash in on business opportunities. Such opportunities abound in the power sector. For example, real time metering and dynamic and differential power tariff for bulk consumers have the potential to substantially improve earnings.

Fifth are the *big seven* approaches to implementation: namely setting stretched goals, cross-functional and spatial coordination, strategic investments, fostering alliances, deploying a deliberative and calibrated approach, and aggressively chasing for results. And finally, but most importantly, there is no substitute for business savvy.

Sixth are *some ground rules for navigating the political economy*. In the democratic context, perception of the electorate matter as much as reality (if not more) and gradualism in reform is essential to mitigate political and social risk as well as for economic prudence. A tripartite strategy has elements of replicability. In implementing a mega transformation, management strategists should carefully consider political positioning, choice of words (phraseology), timing, and a gradual pace for reforms. Incrementalism hedges against risk as reforms are introduced and modified through negotiations with interest groups. In this democratic spirit some battles will be conceded to win the war. And finally, to garner acceptability and credibility for the reforms, deft media management is required.

In conclusion, the Railways's transformation is an exemplar for how state-owned enterprises can improve services despite all the odds of balancing commercial and social objectives to fructify inclusive reforms. While the preceding chapters have unpacked the various attributes of the transformation strategy in substantial depth, this book does not aim to derive prescriptions for reforms. Instead, it aims to provoke the readers by demonstrating that state-owned enterprises that provide infrastructure services can transform quite like private corporations have and that there may be transferable lessons although these may require substantial alteration if applied to alternate sectors, governance structures, and economies because one size never fits all.

Notes

1: Bankruptcy to Billions!

1. One US dollar is 43 rupees at 2008 average market exchange rates. Approximately values have been given in text based on this conversion rate.
2. Operating ratio is the per cent to be spent to earn a unit of revenue—implying that 98 paise were spent to earn a rupee. The ratio is calculated by dividing operating expenditure with operating revenue. The operating expenditure is all cash and non-cash expenses including depreciation and appropriation to pension fund but excluding dividend payable to Government of India. The operating revenue is gross traffic receipts. Therefore, lower the operating ratio, more efficient is the enterprise.
3. Turnaround is the time lapsed between two successive loadings.
4. This echoes the experience of transforming Nissan. 'As long as management gives clear directions that everyone understands, as long as you've got a clear, thoroughly explained strategy, you don't need to worry too much about how well and how fast it's carried out. Don't get me wrong, you still have to expend a lot of time and energy, but it's remarkable how execution falls into place (p. 210).' (Ghosn and Ries 2005: 210)

2: Political Economy of Reforms

1. So every time a passengers peeped out of a moving train they did not have their faces covered with black soot, but as a train pulled out of the station the travellers missed out on the dramatic start of a puffing steam engine—in other words a great deal of the romance of railways.

2. These are also know as the CBC type of coupling and make the train safer and faster. In case of an accident this coupling ensures that coaches remain firmly connected together in the vertical plane. This arrangement prevents one coach from stacking up on another.

3. Earlier, railway track consisted of 90 lb (44.65 kg/m) rails resting on cast iron or wooden sleepers manually laid on a bed of 200 mm size gravel, also known as ballast cushion. In the 1980s and 1990s most of the track structure on the high-density network was upgraded to either 52 kg or 60 kg rail with 72 or 90 ultimate tensile strength, pre-stressed concrete sleepers 1540 to 1660 numbers per kilometre density, and mechanically laid ballast cushions of 250–300 mm size gravel.

4. This is an indicator of the extent to which the cargo customer cross-subsidized the passenger fare. This indicator is arrived at by dividing the average passenger fare per kilometre, by the average freight rate per ton per kilometre.

5. Rail-neer, drinking water bottles produced and sold by Indian Railways, was introduced in 2002.

6. Indian Railways pays a 7 per cent return on this fiscal transfer to the public exchequer. This is a loan in perpetuity.

7. While there is a debate on how reform of infrastructure service providing state-owned enterprises should proceed, privatization and its variant, corporatization, are a significant component of this reform programme. The success of these methods is uneven and varies across geographies as well as sectors.

8. Gauge is the spacing between railway tracks and Indian Railways inherited three track widths—narrow, metre, and broad. Through a gigantic effort, the railways is upgrading the narrower tracks to broad gauge.

9. Despite enormous progress in poverty reduction in India, there are millions of extremely poor people. Based on the 61st National Sample Survey, the Planning Commission, Government of India, estimates the national poverty ratio at 27.5 per cent for 2004. This is substantially lower than the 36 per cent in 1994. But in absolute numbers about 300 million people live below the national poverty line. About 73 per cent of the poor are in rural areas and the rural poverty line is defined as people living below Rs 356.30 for a thirty-day-month (measured by monthly per capita consumption). These rural poor on average have Rs 12 or less to spend each day. The urban equivalent is Rs 538.60 for a thirty-day-month, or Rs 18 per day. In sum, the economic benefits of a growing economy are unevenly distributed across a range of social sub-groups in India (Planning Commission 2007).

10. While this visitor was not from Lalu's political party they belonged to the same social sub-group, or caste, the Yadavs.

11. The car is manufactured by Hindustan Motors and is based on the Morris Oxford model of the 1950s. It is a symbol of indigenously produced automobiles of the post-Independence import substitution and industrialization policies. This car is also know as the Amby and retains its popularity in the corridors of power.

12. See Constitution of India, Articles 74 and 75 on accountability to Parliament and Articles 309–12 on the role of the civil service (Government of India 2003).

13. Price is what consumers will pay for the goods or service, and cost is the sum of the expenses associated with production stages up to the final sale. Among other things, accounting rules about depreciation, amortization, taxes, subsidies, cost of capital, and risks make the reality of a business more complex, but these simple principles remain relevant.

14. Increasing price, or the numerator approach, is preferred when demand is inelastic—the decrease in demand for normal goods is much less than the increase in price, thus increasing total revenue and eventually profitability. The cost-cutting measures, or reducing the denominator, involve reducing costs of inputs, labour, capital, and associated overheads. This approach is preferred for elastic demand—where the consumption of normal goods is very sensitive to prices, especially when substitute goods are available (Hamel and Prahalad 1994: 9). Often a combination is deployed in mixed proportions.

3: THE MARKET

1. The railways's total operational costs consist of fixed and variable costs. Variable costs change with the amount of output produced. Fixed costs, do not vary when the amount of output changes.

2. During this period 95 per cent of total traffic in India was carried by either rail or road transport (Agarwal 2004: 175).

3. This refers to the railways's past practice of price discrimination in freight charges based on the value of the commodity. This practice was also applied in the passenger segments. Such affordability-based price discrimination was benign in the Fabian era because the cost of transportation was borne by the government under the freight-equalization policies. But, post liberalization the cost for freight was no longer a pass through to the government, and producers sought to

minimize costs of freight, further affecting the railways that was already losing market share in steel freight.

4. However, this loss in market share in the steel freight was uneven. For Steel Authority of India the railway did not lose much of its share of freight transport, while with Tata Steel the loss in the freight market share was moderate. In contrast, the recent steel plants of Essar Steel in Gujarat and Ispat Industries in Maharashtra, both in western India, chose to skip the railway services altogether. This differential loss in market share was because of the duality in the type of transportation service railway provides. To deal with the constraints associated with station-to-station freight services of the rail, the old integrated steel plants of Tata Steel and Steel Authority India had developed rail sidings for loading at the factory and unloading at their warehouses in distribution centres. And the railway continued to provide them with a door-to-door service. On the other hand, steel plants like Essar Steel and Ispat Industries directly dispatch steel in trucks to distributors who redirect material to customers. For these new generation steel plants the railways is not competitive in the short or long distance traffic because: the factories dispatch steel in small quantities (less than 2,000 tons), they predominantly serve regional markets (implying short distances, less than 500 km), do not have warehouses to receive large quantities, and the points of consumption are dispersed.

5. These also include commuter trains known by their acronyms DEMU and MEMU (diesel-electric multiple units trains and mainline-electric multiple units). These are engineless trains that can move in both directions.

6. The reason for the vast difference in the proportion of revenues from this relatively small number of mail and express travellers is largely because they travel long distances while the suburban and ordinary passenger travellers travel short distances. This difference is captured in the passenger kilometres. In 2004, the 607 million mail and express train passengers accounted for 292 billion passenger kilometres, while the 4.4 billion suburban and ordinary passenger train travellers accounted for 250 billion passenger kilometres (2005–06 Explanatory Memorandum: Railway Budget, Ministry of Railways: 6). Besides, passenger fares of even unreserved class of mail and express trains are substantially higher than those of ordinary passenger trains.

7. These values include all passenger (Rs 13,298 crore) and other coaching earnings (Rs 923 crore), but exclude negligible sundry earnings.

8. Data for catering and parcel losses in fiscal year 2003 have been used

here because equivalent data for 2004 were not available. In 2004 these figures are likely to have increased.

9. For the purpose of this calculation it has been assumed that all passengers travel on tickets bought for a single trip. In practice many of the suburban rail passengers use discounted monthly season tickets or quarterly season tickets.

10. Here it is assumed that these coaches have full occupancy. Therefore, layout of the coach also determines yield per train.

11. For the purpose of this comparison it is assumed that all coaches accommodate passengers and have full occupancy. The unit cost per train kilometre is calculated based on the methods adopted by Indian Railways—where most cost components are treated as variable, although many of these costs are fixed in nature. Therefore, the cost per train kilometre function increases in a linear manner (for a detailed break-up of fixed and variable costs see Appendix 1).

12. Ratan Tata, the group chairman of Tata Motors, was inspired by the sight of a family of five that can be seen riding on two-wheelers in India. All five ride together, with the father on the front wheel, mother on the pillion, and the children hanging off their parents, like a bunch of grapes. This sparked the reinvention of the people's car (*Economist*, 10 January 2008).

13. Based on data from the Accounts Directorate, Ministry of Railways.

14. This amounts to a loss of Rs 1200 crore.

15. These vans are a safety requirement because Indian Railways does not have anti-collision devices and thus needs these vans as a buffer in the front and rear of the train.

16. Long distance mail and express trains provide opportunities for earnings through parcel and courier services. If a parcel is dispatched from Delhi and delivered in Mumbai in less than twenty-four hours (door-to-door) the market rate for a kg is between Rs 8 and 12. But if the courier time is between twenty-four and fourty-eight hours, then the rates drop to a half. For a courier time of seventy-two hours and more, the rate is as little as Rs 2 per kg. Since mail and express trains are punctual and each train has four brake vans—two in front, two at the rear end, and there is flexibility to add a few more vans—these trains offer an opportunity to capture some of the high-value courier service. The Rajdhani express, departs at 5 pm from Delhi and reaches Mumbai at 10 am (station-to-station), so a parcel couriered via this train can be delivered in less than twenty-four hours (door-to-door). Here the rail is more competitive than roads because buses and trucks (that provide a similar service)

take much longer to travel and have the same additional costs (station-to-door, in their case booking office to customer). On the other hand, for distances less than 500 km, the railway is not competitive for parcel courier because of the station-to-station issues discussed earlier. Further, the railway is not competitive on routes where trains are not punctual, because in the courier market the value is of speed as well as reliability.

17. The railways's parcel service had a three-tiered tariff structure, in increasing order for ordinary passenger trains, mail and express trains, and for the 'super-fast' Rajdhani and Shatabdi trains.

18. (Unit cost = total cost/total output (GTKM) = (fixed cost + variable cost)/total output. As total output ($\uparrow$), fixed costs ($\leftrightarrow$) and variable costs ($\uparrow$), therefore unit cost ($\downarrow\downarrow$).

19. In 1991 the unit cost per net ton kilometre of the Indian Railways was 10 paise which has reduced to 4.3 paise in 2004 at constant prices (base year 1982). While in 1981, the railway had 11,000 engines by 2004 the railway had a fleet of 7800 engines—predominantly diesel and electric powered. Likewise, the 227,000 wagons that are part of the railway's rolling stock are half of the past quantities. But, the engines and the wagons are superior in technology and have substantially increased the railways's productivity. In addition, from 1981 to 2004, the number of employees had declined from 15.5 million to 14.4 million.

20. In nominal terms the railway's costs were increasing faster than its revenues, implying an inevitable financial crisis.

21. Also known as GTKM, it is a combine measure that captures the distance and weight of total traffic carried on the railways.

22. Divided by a weighted input cost index for the railways. Weighted index of inputs for railways (WIIR) is prepared annually in order to measure change (increase or decrease) of railway's input costs over time. The purpose of this index is to compute a single index from price indices of the inputs relevant to the railways, taking into consideration their relative weights in order to produce a composite metric (for details see Appendix 2).

23. Amount of traffic transported in a given time.

24. A measure of the amount of load transported by diesel engines.

25. Statistics and Economics Directorate, Ministry of Railways, Government of India.

26. Operating leverage is the elasticity of profit with respect to output.

27. In parenthesis is the percentage change in values with respect to the values under seventeen coaches.

28. Here the assumption is that additional coaches have full occupancy and the engines can haul these longer trains. In response to such needs Indian Railways had acquired high-horsepower locomotives but could not increase the length of the trains because the platforms at railway stations were not long enough to accommodate longer trains.

29. The cost per passenger kilometre is 41 paise and the average weight of a passenger is assumed to be 60 kg (17 passengers × 60 kg = 1 ton. Therefore, 17 travellers × 41 paise = 697 paise per ton kilometre).

30. This case study train accommodates more passengers in each coach (consider a chair-car coach instead of a sleeper coach). While the Rajdhani's coach carries forty-seven travellers, each coach in the case study train carries about eighty passengers.

31. Profit $(\pi) \neq f$ (fares); instead Profit $(\pi) = f$ (fares + length of train + occupation rate in coaches + speed of train + weight transported per wagon + composition of rake).

    All data is from the Statistics and Economic Directorate, Ministry of Railways, Government of India, New Delhi.

4: Milking the Cow

1. The budgeted target for fiscal year 2009 is 850 million tons and the internal target is of 870 million tons. On achieving the budgetary target this year, the incremental loading over the five years (2005–09) will be 293 million tons.

2. The design load-bearing capacity of each of the four axles of a wagon.

3. For several reasons Indian Railways's freight trains are not comparable, yet there was a lot of scope for improvement. Australia and Brazil were the frontrunners, carrying 44,000 tons and 30,000 tons respectively. Chinese railway was hauling trainloads of 200,000 tons and railways in Russia, South Africa, and the United States, ran trainloads of 15,000 tons each (UIC 2007).

4. Average distance travelled by a train—known as lead in rail parlance— was 660 km in 2004. As most trains returned empty and the average speed of a freight train was 24 km per hour, of the seven days only 2.3 days, or fifty-five hours, were spent on the move.

5. This is because increasing the speed of trains requires complex interventions, many of which can only be achieved with massive investment in new tracks and signalling technology, that too in the long term.

6. The final decision was left to the zonal railways so as to consider local constraints. Authority was devolved to general managers at zonal level to make adjustments as per local conditions. In some cases, land for extending platforms was not available, while in others, workers' unions were resistant. Further, the extension of platforms was accompanied by improved illumination, construction of access roads, provision of electricity back-up generators, and provision of other infrastructure.

7. This policy instrument sets the priority in which trains will be received at a terminal for loading and unloading or determines the right of way. Customers are ranked into five categories in the following order— defence receives the first right of way, followed by relief and emergency freight services, programmed traffic for feeding power plants and steel factories, followed by cement and other customers.

8. The examination was valid for 4500 km and the average round trip was 1320 km.

9. Many train examination points located at private terminals have been closed, validity of closed-circuit brake-power certificates have been increased to thity-five days and a new concept of premium end-to-end train examination with BPC (brake power certificate) validity of fifteen days has been introduced. Out turn from shops and sick lines has shown improvement on account of close and regular monitoring against specified targets.

10. The light-weight-weaker rails—90R 72 ultimate tensile strength— were replaced with heavier and stronger rails—60 kg with 90 ultimate tensile strength. Further, the railway had upgraded wooden and cast iron sleepers with higher quality pre-stressed concrete sleepers. But upgrading of tracks was done in a piecemeal manner, requiring entire routes to operate on the basis of older tracks. As recent as 2004, the high density Delhi–Mumbai and Delhi–Kolkata routes had its track segments built of the weaker and older rail, reconciling the trains on these routes to function once again on the lowest common denomination.

11. As criteria for replacement of old tracks was based on the age of assets, several intermittent segments were waiting their time out. However, due to this bottleneck, the entire new investment did not yield improved productivity.

12. XE and XG class engines.

13. This condition was compounded by the fact that the track modulus— the physics equations to calculate the permissible load on tracks—had

not been revised despite significant improvements in the quality of tracks.

14. Association of American Railroads definition: US Class I Railroads are line haul freight railroads with 2006 operating revenue in excess of US$ 346.8 million. http://www.aar.org/~/media/AAR/Industry%20Info/Statistics.ashx

15. Some illustrative 'aspects of resource leveraging' are summarized as converging, focusing, targeting, learning, borrowing, blending, balancing, recycling, co-opting, protecting, and expediting (p. 191).

16. Doubling refers to laying a second railway track along an existing route. Likewise, a third track is referred to as tripling and so on.

17. The railway's total rolling stock is worth US$ 25 billion and the capital invested in wagons is a whopping Rs 30,000 crore, (US$ 7 billion), locomotives worth Rs 50,000 crore, (US$ 12 billion) and coaches worth another Rs 40,000 crore (US$ 9 billion). But, in the past investment focused on acquiring new assets: Rs 11,000 crore was spent acquiring new assets in 2005. While in the same year, Indian Railways spent a paltry Rs 14 crore (US$ 3.3 million) on capital works for upgrading infrastructure for maintaining these wagons.

18. New policy initiatives—increasing axle load, rationalizing train examination practices, round-the-clock working at freight terminals, route-wise planning, traction change points, incremental revenue, and the like—require a chase tool-kit. In the case of axle load increase, first a stretched target was set—mission 600 million tons, internal target for 2006. However, the issue of enhancing axle load began with the Minister's surprise inspection at Danapur where overloading was identified. As a follow-up of that visit and the related fall-out, the Minister issued a memorandum. This first note was personally signed by the 'MR' as the Minister for Railways is known. In the note, not only were priority issues identified, but the Board was also instructed to respond within a specific time period of two weeks. Further, to chase the issue, as the Minister's office sought past research on axle load and this was forwarded to the Railway Board, and each supplementary note acted as a reminder. For instance, the office of the Minister forwarded, supplementary notes on the 22 and 23 tons axle load steam locomotives that operated in 1922 on weaker tracks. This was followed by CANAC Incorporated, Asian Development Bank, and International Heavy Haul Association's assessments on clearance of the railway tracks for 29 and 30 tons axle load. Still the file on axle load was not returned to the Minister. Therefore, a follow-up meeting with the minister was

convened, where the agenda required an update on the progress made on increasing axle load.

Each individual member of the taskforce on increasing axle load was contacted for routine updates on progress. Feedback from multiple members provided a sense of the debate and reasons for delay. Irrespective of hierarchy, staff inputs were solicited. In many instances junior staff were most receptive to provide feedback on the obstacles and departmental concerns. Keeping in mind the delicate relationships within the departmental hierarchy, this feedback was deployed to nag the departmental heads. If the axle load issues were resolved by the time the meeting with the Minister was scheduled, then the meeting would be cancelled. However, in this case the meeting was necessary. On the day of the meeting the Minister was provided a comprehensive multi-departmental view on the obstacles to increasing axle load. And the meeting with the Minister broke the deadlock and the first decision on increasing axle load by 2 tons on one route on one train was taken. While the initial policy note had thirty-four routes under consideration, the approval of two routes was a success. As soon as these two routes were approved, the next file chased the remaining thirty-two routes. After relentless pursuit, and forty-five circulars on axle load, most routes carrying heavy bulk commodities allow freight trains load to carrying capacity plus 6 tons per wagon. But, in other cases this was not enough. This approach of management by nagging will be discussed in some detail in the chapter on management strategies.

19. Net result, the improved BCN and BOXN rakes will carry a payload of 4100 tons, an increase of 78 per cent from the previous 2,300 tons. In some cases, the train length has been extended from forty to fifty-eight wagons, axle load has been increased from 20.3 tons to 22.9 tons and tare weight reduced by 4 tons. For BOXN wagons the capacity had been enhanced by 22 per cent to 4100 tons.

20. The recent spike in inflation and interest rates is a new trend and is affecting the unit costs this fiscal year.

5: Service with a Smile

1. Both these services have certain temporal, spatial, and other conditions that circumscribe the benefits of these new products to the customers so that the railways needs to offer added value in order to retain or regain.

2. Since the railways monitors the daily loading of cement and steel, the decline was even more obvious.

3. The reformers had deliberated on these issues earlier, and the Railway Board had approved modifications in the policy. Working with consensus was essential to functioning in a mega bureaucracy.

4. Further, commodities in class 200 had three different weight conditions and class 140 had four.

5. Post-Independence, the railways carried most long-distance freight because the road sector was in its nascent stage.

6. In 1995, acceptance of smalls was formally terminated and almost all freight migrated to train loads.

7. ABC analysis prioritizes items in a rank order based on their contribution to the total net-ton-kilometre. The results were grouped into three bands—namely A, B, and C—where category A consisted of eight commodities that contributed to over 85 per cent of total freight volumes, category B included an additional sixty-three commodities such that A and B categories combined accounted for 97 per cent of total freight volumes, and category C accounted for the remaining 3 per cent.

8. Any other alloy and metal that has not been listed will be charged the same class freight rate as the generic class—metals and alloys. In addition to the twenty-four group heads, four heads have been created for light-weight commodities and account for a negligible proportion of the total freight, but since they are in use they have been accommodated. Because the objective of this tariff rationalization was to simplify the cumbersome routine, and not to increase revenue, four additional block classes below class 90 were introduced so as to minimize the increase in effective freight charges for these commodities.

9. The rate table provides a matrix that varies prices by rate class and for a range of distances.

10. In essence dynamics of pricing are embedded in the micro conditions and depend on the origin and the destination, seasons, fuel prices, the number of service providers, condition of the roads, and a host of other variables that affect demand and supply. However, Indian Railways was detached from this market dynamics.

11. The floor price was set because loading and unloading take a day each and there is an incremental cost to hauling loaded trains as they travel slower and consume more energy and these costs need to be recovered.

12. While in the monsoon season, which is the lean season, freight trains are stranded as tracks and mines get flooded, less coal is consumed by

power plants because hydroelectric power plants come online, and drop in construction activity, dampens demand for steel and cement.

13. Profit margin is the difference between unit price and unit cost. In the context of the railway price as well as costs differ by product segments— based on quality and quantity of service provided and the demand in the market. To increase margins either price needs to increased, or cost needs to be decreased, or both.

14. Another paradox is that while freight earnings from seven main commodities other than iron ore increased by 9 per cent the overall earnings in freight increased by 23 per cent. While the freight volume, transported as measured in net-ton-kilometres, of iron ore for export is 6 per cent of the total freight volume in the fiscal year 2008, it accounted for 9 per cent of the total freight revenue. In the current fiscal year, that is 2009, while the iron ore for export retains its share of total freight volume at 6 per cent, its share of total revenue is 20 per cent. While last year's margin was 100 per cent, this years' margin has leaped to a whopping 250 per cent. This year iron ore accounts for 13 per cent of the growth in the first four months.

15. Unlike the past, where up to 40,000 indents would stack up waiting for the rakes because demand far exceeded supply, through supply-side management, demand for rakes is met year round except during the peak season where a congestion charge has been introduced in the from of a peak-season surcharge.

6: Outcomes, Sustainability, and Replication

1. But similar efforts have been less successful in other contexts, like Mexico for instance.

References

Agarwal, V. (2004). *Managing Indian Railways*. New Delhi: Manas Publications.

Bhandari, R.R. (2006). *Indian Railways: Glorious 150 Years*. Second edition. New Delhi: Ministry of Information and Broadcasting, Government of India.

Budget Speech. (February 2002). Railway Budget 2002–03. Ministry of Railways, Government of India. New Delhi, May.

————. (February 2004). Railway Budget 2004–05. New Delhi: Ministry of Railways, Government of India.

————. (24 and 26 February 2005). Railway Budget 2005–06. New Delhi: Railway Budget 2005–06, Ministry of Railways, Government of India.

————. (24 February 2006). Railway Budget 2006–07. New Delhi: Ministry of Railways, Government of India.

————. (24 and 26 February 2007). Railway Budget 2007–08. New Delhi: Ministry of Railways, Government of India.

————. (2008a). Indian Railways Key Statistics 2007–08. New Delhi: Directorate of Statistics and Economics, Ministry of Railways, Government of India.

————. (2008b). *Data Book, Railway Budget (2005–06 and 2006–07)*. New Delhi: Ministry of Railways, Government of India.

————. (26 February 2008). Railway Budget 2008–09. New Delhi: Ministry of Railways, Government of India.

Business Standard. (27 February 2005). *So whose Railway Budget was it?* [Aditi Phadnis, Politics section].

Business Week. (28 November 2005). *The Man Who Invented Management: Why Peter Drucker's ideas still matter* [Cover story by John Byrne]. *Business Week*, The McGraw-Hills.

Economist. (10 January 2008). 'A"people's car" from India: Tata Motors reveals a dirt-cheap model,' London.

Ethos of Indian Railways, Tandon Committee Report. New Delhi: Ministry of Railways, Government of India, March.

Frontline. (2001). 'On the Wrong Rack'. Vol. 18, Issue 14, July 7–20.

Ghosn, C. and P. Ries. (2005). *Shift: Inside Nissan's Historic Revival*. New York: Doubleday.

Government of India. (2003). *The Constitution of India*. New Delhi: Ministry of Law and Justice, Government of India.

Greenspan, A. (2007). *The Age of Turbulence: Adventures in a New World*. London: Penguin.

Hamel, G. and C. Prahalad. (1994). *Competing for The Future*. Boston, Massachusetts: Harvard Business School Press.

Harral C., Sondhi, J. and Chen, G. World Bank. (2006). 'Highway and Railway Development in India and China, 1992–2002'. *Transportation Notes, Roads, Highways and Rural Transport Thematic Group*. Washington DC: The World Bank Group.

India Today. (29 May 2006). The Best And The Worst [cover story: ministers' ranking by Shankkar Aiyar]. New Delhi: *India Today*.

International Union of Railways. (2006). *Railway Statistics—Synopsis*. Paris: International Union of Railways. Web link: http://www.uic.asso.fr/?changeLang=en

Ministry of Railways. (May 2002). *Status Paper on Indian Railways: Issues and Options*. New Delhi: Ministry of Railways, Government of India.

————. (2008a). *Indian Railways Key Statistics 2007–08*. New Delhi: Directorate of Statistics and Economics, Ministry of Railways, Government of India.

————. (2008b). *Year Book (1987–88 to 2007–08)*. New Delhi: Ministry of Railways, Government of India.

————. (2008c). *Annual Reports and Accounts (1987–88 to 2007–08)*. New Delhi: Ministry of Railways, Government of India.

————. (2008d). *Annual Statistical Statements (1987–88 to 2007–08)*. New Delhi: Ministry of Railways, Government of India.

————. (2008e). *Data Book, Railway Budget (1987–88 to 2007–08)*. New Delhi: Ministry of Railways, Government of India.

Mohan, R. (2001a). *The Indian Railways Report 2001: Policy Imperatives for Reinvention and Growth*. Vol. 1: Executive Summary. New Delhi: Expert Group on Indian Railways.

——. (2001b). *The Indian Railways Report 2001: Policy Imperatives for Reinvention and Growth.* Vol. 2, Part 1. New Delhi: Expert Group on Indian Railways.

Pandya, M. and R. Shell. (2005). *Lasting Leadership: What Can You Learn from the Top 25 Business People of Our Times.* Upper Saddle River, New Jersey: Wharton School Publishing.

Planning Commission. (1988). *Report of the Steering Committee on Perspective Planning for Transport Development.* New Delhi: Planning Commission, Government of India.

——. (2007). 'Poverty Estimates for 2004–05'. New Delhi: Ministry of Railways, Government of India.

Raghuram, G. and N. Shukla. (2007). 'Turnaround' of Indian Railways: Increasing the Axle Loading. Working Paper Number 2007–05–03. Ahmadabad: Indian Institute of Management, May.

Railway Gazette. (2008). Rio Tinto to go driverless. *Railway Gazette, International.* June19.Weblink: http://www.railwaygazette.com/news_view/article/2008/06/8554/rio_tinto_to_go_driverless.html

Reliance Communication. (2008). *A Fuller Biography of Dhirubhai H. Ambani.* Retrieved 9 October from: http://www.rcom.co.in/wobapp/Communications/rcom/Aboutus/about_da.jsp

RITES. (October 1997). *Study on Decline in Railway's Share in Total Land Traffic in India.* Vol. 1. New Delhi: Rail India Technical and Economic Services (RITES), Ministry of Railways, Government of India.

Slater, R. (2003). *29 Leadership Secrets from Jack Welch.* New York: McGraw-Hill.

Sondhi, J. (2002). 'Railways.' Chapter 3 in *India's Transport Sector: The Challenges Ahead.* Vol. 2: Background Papers, 10 May. Washington DC: The World Bank Group.

Tandon, P. (1994). *Organizational Structure and Management Ethos of Indian Railways,* New Delhi: Tandon Committee Report, Ministry of Railways, Government of India, March.

Thompson, L. (2003). 'Changing Railway Structure and Ownership: Is Anything Working?' *Transport Reviews,* Vol. 23, No. 3, 311–55.

TRAI. (2007). 'A Journey Towards Excellence in Telecommunications.' New Delhi: Telecom Regulatory Authority of India (TRAI).

——. (2008). 'The Indian Telecom Services: Performance Indicators October–December 2007'. 10 April. Web-link: http://www.trai.gov.in/trai/upload/Reports/41/preport10april08.pdf

UIC. (2006). *International Railway Statistics 2006 [UIC Leaflet]*. Paris: International Union of Railways (UIC)

———. (2007). *International Railway Statistics 2007 [UIC Leaflet]*. Paris: International Union of Railways (UIC).

Van der Meulen, D. and F. Möller. (2006). 'Railway Globalization: Leveraging Insight from Developed into Developing Regions'. *World Congress on Railway Research*, 4–8 June. Montreal. Web link: http://www.railcorpstrat.com

Welch, J. and J. Byrne. (2001). *Jack: Straight from the Gut*. New York: Warner Books.

World Bank. (2006). 'Highway and Railway Development in India and China, 1992–2002.' *Transportation Notes, Roads, Highways and Rural Transport Thematic Group*. Washington DC: The World Bank Group, May.

Glossary

HINDI WORDS

Aam admi	common man
Bhajiya	deep fried snacks
Coolie	a porter
Khadi	hand spun and hand woven cotton cloth
Kulhads	clay pots
Mumbaikars	the people of Mumbai
Rail Bhavan	headquarters of the Ministry of Railways, Government of India
Raj	rule
Yatri niwas	hotels

TECHNICAL TERMS

Zonal Railways	Indian Railways is divided into sixteen zonal railways. Each zone has a territorial jurisdiction ranging from 2400 to 6500 route kilometres. Each zone is led by a General Manager (GM).
Divisions	Each zonal railway is divided into three to six operational units called divisions. There are 68 divisions in Indian Railways each led by a Divisional Railway Manager (DRM).
Production Units	Indian Railways has six production units for manufacture of coaches, locomotives, and the like.

Passenger Earnings	Earnings from all classes of passenger traffic.
Other Coaching Earnings	Earnings from luggage, parcel, postal, defence, and other miscellaneous coaching services.
Goods Earnings	Earnings from all kinds of goods traffic.
Sundry Earnings	Earnings from leasing of land, advertising, catering, right of way, dividend from public sector undertakings, reimbursement of operating loss on strategic lines, rents, tolls, and the like.
Total Earnings	Total earnings from goods, passenger, other coaching, and sundries.
Suspense	Traffic earnings which have accrued but not yet realized.
Gross Traffic Receipts	Earnings realized, namely total earnings +/- suspense.
Miscellaneous Receipts	Receipts from subsidy received from general revenues towards dividend relief, receipts from railway recruitment boards, and the like.
Total Receipts	Sum of gross traffic receipts and miscellaneous receipts.
Ordinary Working Expenses	Expenses incurred for operation of the railway system including general superintendence and services; repairs and maintenance of permanent way and works, motive power and rolling stock, plant and equipments; traffic expenses, fuel, staff welfare, and the like.
Appropriation to Pension Fund	Amount appropriated from total receipts to the pension fund.
Appropriation to Depreciation Reserve Fund	Amount appropriated from total receipts to depreciation reserve fund.
Total Working Expenses	Ordinary working expenses (+) appropriation to pension fund (+) appropriation to depreciation reserve fund.
Open Line Works Revenue	Small works charged to miscellaneous expenditure.

Miscellaneous Expenditure	Expenditure on Railway Board and other offices like RDSO, training institutes, open line works revenue, and the like.
Total Expenditure	Total working expenses (+) miscellaneous expenditure.
Net Revenue	Total receipts (-) total expenditure.
Dividend Payment	It is the amount of dividend paid by the Ministry of Railways to Ministry of Finance for the capital investment made by it in the railways.
Surplus	Net revenues (-) dividend payment.
Cash Surplus before Dividend	Total receipts (+) interest on railway fund balances (+) appropriation to depreciation reserve fund (+) open line works revenue.
Investible Surplus	Cash surplus before dividend (-) dividend payment (-) interest on pension fund.
Appropriation to Development Fund	Amount appropriated to development fund from surplus.
Appropriation to Capital Fund	Amount appropriated to capital fund from surplus after appropriation to development fund.
Depreciation Reserve Fund	This fund provides for expenditure on replacement and renewal of assets including improvement element.
Pension Fund	This fund provides for the expenditure on pensionary charges of railway employees.
Development Fund	This fund provides for expenditure on providing amenities for passengers and other railway users, labour welfare works, un-remunerative operating improvements, and the like.
Capital Fund	Provides for capital expenditure.
Budgetary Support	Funds received from General Exchequer for capital expenditure.
Capital-at-Charge	Book value of railway assets created from budgetary support.
Return on Capital Investment	Rate of return on capital investment made on the Railways. Ratio = (net revenue)/(capital-at-charge (+) investment from capital fund).

Operating Ratio	Ratio of total working expenses excluding suspense to gross traffic earnings.
Traffic Revenue Earning	Traffic which is paid for by the consignor or the consignee.
Gauge	Indian Railways is a multi-gauge system, comprising of broad gauge (1676 millimetre), meter gauge (1000 millimetre), and narrow gauge (762 or 610 millimetre).
Key-men	Railway employee who inspect their designated stretch of rail network by foot, both track and bridges, once a day.
Private Sidings	Refers to the length of line from the take off point on the main railway system to the siding holder's premises.
Piecemeal or Smalls	Goods consignments whose weight and dimensions do not require the exclusive use of a wagon.
Gross ton kilometre	A ton, including payload, tare and weight of engine, carried over one kilometre.
Net ton kilometre	Payload of one ton carried over one kilometre.
Wagon Turnaround	Interval of time between two successive loadings of a wagon.
Passenger kilometre	A passenger transported over one kilometre.
Train kilometre	Movement of a train over one kilometre.
Lead	Average haul of a passenger or a ton of freight.
Output	The volume of traffic moving between any two points on the railways expressed in terms of passenger kilometres (PKM) or net ton kilometres (NTKM) or train kilometers per running track kilometre.
Sick Line	A portion of the yard where carriage and wagon repairs are undertaken.
Pit Line	A maintenance line for under-frame examination of rolling stock.
Yard	A portion of the railway system where trains are dealt with.
Workshop	A facility for repair of rolling stock.

Wagon	A freight truck used for carrying goods in railways.
S and T	Signalling and telecommunication used for controlling the movement of trains.
Free Time	Time limit prescribed for free of charge loading/ unloading and removal of consignments from railway premises.
Train Examination	Examination of rolling stock at fixed intervals.
Rake	A formation consisting of a set of coaches/wagons to be hauled as a train.
Close-circuit Rake	To and fro movement of goods trains over a pre-defined route with a base depot for maintenance and examination.
Mileage based Examination	Maintenance schedule undertaken after movement of a train over a pre-determined distance.
End-to-end Rake	Rakes which are offered for examination and maintenance before every loading.
Passenger Profile Management	Mapping of reservation demand against availability of accommodation with a view to optimize the latter.
Rake Links	Deployment of passenger trains between originating and destination stations with a view to optimize utilization of rolling stock.
Axle load	The axle load is the maximum weight of an axle permitted on a given track and is a governing parameter for rolling stock and track design. It is calculated by dividing the gross weight of loaded freight wagon/coach/locomotive by number of axles of the rolling stock.
Electronic Weigh Bridge	A device used for weighing each loaded wagon of the train electronically.
Wagon Impact Load Detector (WILD)	Real-time online system to capture, analyse, and report the instantaneous loads exerted by the passing train wheels on the instrumented track section. The system is capable of isolating and identifying wheels exerting excess impact load on the track and send this report immediately to the Divisional Control through an electronic network.

Wharfage	Charges accrued on use of railway premises beyond free time.
Throughput Enhancement	Throughput enhancement refers to any investment or activity that results in an increase in the capacity of a particular section or route to carry additional traffic measured in terms of gross ton kilometres.
Route-wise Planning	Route-wise planning refers to an approach whereby integrated planning is done for the entire route so that bottlenecks are eliminated completely leading to a significant increase in output as against piecemeal/incremental planning which often result in relocating, not fixing, the bottlenecks.
High Density Network (HDN)	This comprises of the rail routes connecting the four metros of Delhi, Kolkata, Chennai, and Mumbai with each other and the Delhi-Guwahati route. HDN carries bulk of the freight and passenger traffic of Indian Railways.
Double Line Track	Two separate unidirectional lines running in parallel.
Block Section	The portion of running line between two stations.
IBS	An arrangement of signalling in which a long block section is split into two block sections for enhancing line capacity.
Multimodal Logistic Park	An inter-model cargo landing and processing facility consisting of container terminals, warehouses, banks, office space, and mechanized facilities for landing, sorting/grading, cold chain aggregation/disaggregation, and the like. The objective of this facility is to provide an efficient and integrated logistics hub in or near special economic zones, industrial clusters, or other nodes of cargo convergence.

Appendices

Appendix 1

Assumptions for cost allocations are as follows. The costs are split into variable and fixed costs based on Railways's cost allocation principles. The variable costs are discussed, followed by fixed costs.

First, unit cost of terminal expenses per passenger originating is applied and as such it is treated as a variable cost, although many of these costs are fixed in nature. Terminal expenses include the expenses relating to ticket booking office, ticket checking, reservation and enquiry, special services such as expenses relating to retiring rooms, waiting rooms, platform inspectors, passenger guide, and travelling ticket checking staff. Additionally, costs for miscellaneous services include expenses incurred on provision and maintenance of platforms, washing, cleaning, and water provision for coaches, train examinations, and shunting activities.

Second, unit cost per coach kilometre (number of coaches multiplied by the kilometres travelled) is applied to compute line haul cost of traction—the unit cost for hauling a train for one kilometre and thus varies with the number of coaches—includes cost of fuel, maintenance cost of overhead electric traction, repair and maintenance of locomotives, and related depreciation, interest as well as cost of crew, therefore it's a variable cost.

Third, the line haul cost of track maintenance is considered a variable cost. While many of these costs are fixed in nature but are treated as variable because they are computed based on the unit cost per coach kilometre. It includes all the repairs and maintenance cost of tracks, depreciation and interest on tracks.

There are two types of costs that are treated as fixed where addition or removal of coaches does not change costs on unit distance travelled. Therefore, unit cost per train kilometre is a fixed cost. Line haul cost of transportation includes cost of maintenance of structures (other than track), operating staff, train passing staff, and the cost of signalling are computed based on the unit cost per train kilometre of these activities.

To account for overheads, central charges are added as a percentage of total direct costs and therefore they too vary with the number of coaches.

Further, repairs and maintenance, depreciation and interest of coaches vary directly in proportion to the number of coaches and are calculated on travel time of the train based on the capital cost of coaches.

As a result of the application of the above unit cost adopted by the Indian Railways—where most costs components are treated as variable, although many of these costs are fixed in nature—the cost per train kilometre function increases in a linear manner.

Appendix 2

Weighted index of inputs for railways (WIIR) is prepared annually in order to measure change (increase or decrease) of railway's input costs over the time. The purpose of this index is to compute a single index from price indices of the inputs relevant to the railways, taking into consideration their relative weights in order to produce a composite metric.

The wholesale price indices (WPI) consider a general basket of commodities, composed of (a) primary articles—food and agricultural products; (b) fuel, power, light and lubricants; (c) manufacturers products. Many of these articles are not relevant for the Railways's operations. Therefore, railway specific inputs constitute the WIIR—diesel, electricity, transport equipment, non-ferrous metals, electrical machinery, lubricants, oil, manufactured products, heavy Rails and ferrous metals. Further, the WPI has indices for these constituents, and are used to derive the WIIR. Therefore, the base years of these constituents correspond to the base years of WPI. For the purpose of computing real prices for the railways, inflation adjustments are made based on this railway specific WIIR and not the general WPI.

Source: Statistics and Economics Directorate, Ministry of Railways, Government of India, 2008.

Appendix 3

Plan expenditure over the years

(All amounts in Rs crore)

	Total	General exchequer	Internal resources	External resources
2001	9395	3269	3229	2897
2002	10177	5517	2485	2175
2003	11408	5778	3113	2517
2004	13394	7081	3476	2837
2005	15422	8669	3712	3041
2006	18838	8073	7034	3731
2007	25002	7914	12233	4855
2008	28680	8668	14948	5064

Appendix 4

Rate circular number & date	On routes applicable
CC+2	
RC. No. 22 of 2004 w.e.f 1.9.2004	All loose and bulk commodities
RC. No. 31 of 2004 dated 16.09.2004	Consolidation of all earlier circulars for chargeable weight for CC commodities
CC+4	
RC. No. 48 of 2004 dated 7.11.2004	For all eight-wheeled wagons for loose and bulk commodities
CC+6	
RC. No. 67 of 2005 dated 21.1.2005	6 routes of CC + 6 were introduced
RC. No. 11 of 2006 dated 01.02.2006	Some new routes added in CC + 6 routes
RC. No. 15 of 2006 dated 17.02.2006	New routes added
RC. No. 27 of 2006 dated 29.03.2006	New routes added
RC. No. 41 of 2006	Compilation of CC+6 routes
RC. No. 82 of 2006 dated 5.10.2006	10 new CC+6 routes are added from different railways
RC. No. 96 of 2006 dated 13.11.2006	4 new CC+6 routes are added
RC. No. 103 of 2006 dated 30.11.2006	5 new CC+6 routes are added
RC. No. 10 of 2007 dated 07.02.2007	17 new CC+6 routes added
RC. No.16 of 2007 dated 28.02.2007	1 new CC+6 routes added
RC. No. 69 of 2007 dated 27.06.2007	Consolidation of all CC+6 circulars issued till date

Appendix 4 (*Contd*)

Rate circular number & date	On routes applicable
RC. No. 77 of 2007 dated 24.07.2007	11 new CC+6 routes added
Add-II dated 05/07/2007	2 new CC+6 routes added
RC. No. 76 of 2007 dated 20.07.2007	All routes except a few have been notified as universalized CC+6 routes effective from 06.08.2007
CC+8	
RC. No. 25 of 2005 dated 10.05.2005 effective from 15.05.2005	16 CC+8 routes identified
RC. No. 29 of 2005 dated 2.06.2005	4 more CC+8 routes added
RC. No. 42 of 2005 date 13.07.2005	1 more CC+8 route added
RC. No. 45 of 2005 of dated 19.07.2005	2 new CC+8 routes added effective from 22.07.2005
RC. No. 73 of 2005 dated 19.12.2005	2 new CC+8 routes added effective from 01.01.2006
RC. No. 76 of 2005 dated 29.12.2005	3 new routes in CC+8 are added effective from 15.01.2005
RC. No. 10 of 2006 dated 01.02.2006	3 new routes added in CC+8
RC. No. 41 of 2006 dated 10.05.2006	Compilation of all CC+8 routes
RC. No. 51 of 2006 dated 08.06.2006	Sugar, rice, cement in BCNA/ BCNHS as 64 tons and 61 tons in BCN on CC+8/CC+6 routes
RC. No. 97 of 2006 dated 1.11.2006	When traffic moved comparing on CC+8/CC+6 and normal routes- PCC of wagons
RC. No. 48 of 2007 dated 26.04.2007	1 CC+8 route added
RC. No. 54 of 2007 dated 18.05.2007	1 CC+8 route added
RC. No. 69 of 2007 dated 27.06.2007	Consolidation of all CC+ 6/ CC+8 circulars
Addendum No. 3 of RC. No. 69 dated 10.07.2007	2 CC+8 routes added
Addendum No. 4 of RC. No. 69 dated 10.09.2007	5 CC+8 routes added
Addendum No. 5 of RC. No. 69 dated 21.09.2007	2 CC+8 routes added
Addendum No. 6 of RC. No. 69 dated 22.11.2007	8 CC+8 routes added
Addendum No. 7 of RC. No. 69 dated 30.11.2007	25 new CC+8 routes added

Addendum No. 9 of RC. No. 69 dated 04.01.2008	24 new CC+8 routes added
Addendum No. 10 of RC. No. 69 dated 08.04.2008	9 new CC+8 routes added
Addendum No. 11 of RC. No. 69 dated 08.04.2008	1 new CC+8 route added
RC. No. 77 of 2007 dated 24.07.2007	Slag, E&F Grade Coal permitted to be loaded up to CC+8 i.e. 67 tones in BOXN/ BOXNHS on CC+8 routes
RC. No. 102 of 2007 dated 6.11.2007	PCC of BOXNCR and BOXNHA on CC+6/CC+8 routes have been notified.
RC. No. 28 of 2008 dated 18.06.2008	Compilation of all CC+8 routes circulated
Addendum No. 1 of RC. No. 28 of 2008 dated 20.06.2008	1 new route included which was left out while compiling
Addendum No. 92 of RC. No. 28 of 2008 dated 08.07.2008	2 more CC+8 routes added

Appendix 5

Radical Incrementalism

Minimum weight condition (MWC) for different commodities till 1987; Introduction of BOXN wagon (1983); MWC of Slack Coal—Marked 55 tons in BOXN wagon (1987); Slack Coal notified to charged at CC in BOXN wagons (1997); ROM Coal and Iron & Steel are notified to charges at CC+2 in BOXN wagon (1998); On KK Line—Iron Ore in BOXN wagon is being charged at CC+4 (1986); In South Eastern Railway, Local Iron is being charged at CC+2 (1987); In Korba and IB Valley Slack Coal is being charged at CC+2 (1990); In South Eastern Railway Iron Ore Dolomite, Limestone, Manganese Ore, Powerhouse Coal, and Washed Coal are notified to be charged at CC+4 (for 1st February to 31st March 2004).

Appendix 6

Statement of Cash and Investible Surplus

The Railways's accounting method was revised between 2005 and 2008. For the data to be comparable all figures presented below (2001–08) are recomputed based on the prevalent accounting method.

All amounts in crore rupees

Description	2001	2002	2003	2004	2005	2006	2007	2008
1. Ordinary Working Expenses	26,449.35	27,586.44	28,407.87	29,311.53	31,718.45	35,029.53	37,432.53	41,033.17
2. Appropriation to Pension Fund	4841.85	5600.00	5950.00	6263.09	6680.00	6950.00	7426.00	7989.00
3. Miscellaneous Expenditure	227.12	230.44	244.66	275.11	287.44	319.11	407.45	423.59
4. Total [1 to 3]	31,518.32	33,416.88	34,602.53	35,849.73	38,685.89	42,298.64	45,265.98	49,445.76
5. Gross Traffic Receipts	35,154.31	38,140.71	41,377.23	43,143.62	47,661.58	54,773.29	63,040.50	71,720.06
6. Miscellaneous Receipts	1130.47	1520.22	1673.25	2005.68	1676.37	1824.13	2054.34	1556.51
7. Total Receipts [5+6]	36,284.78	39,660.93	43,050.48	45,149.30	49,337.95	56,597.42	65,094.84	73,276.57
8. Interest on Railway Fund Balances	23.04	42.53	135.30	252.70	392.17	580.35	818.63	1,175.38
9. Cash Surplus before Dividend [(7+8)–4]	4789.50	6286.58	8583.25	9552.27	11,044.23	14,879.13	20,647.49	25,006.19
10. (a) Dividend Payable to General Revenues[1]	581.47	1640.30	2973.84	3325.76	3007.38	3286.83	3892.81	4238.93

(b) Payment of Deferred Dividend	..	..	50.00	300.00	483.30	663.00	663.00	664.00
Total Dividend Payment	581.47	1640.30	3023.84	3625.76	3490.68	3949.83	4555.81	4902.93
11. Cash Surplus after Dividend [9–10]	4208.03	4646.28	5559.41	5926.51	7553.55	10,929.30	16,091.68	20,103.26
12. Appropriation to D.R.F. along with interest	2305.43	2024.49	2482.11	2742.83	2893.29	3843.06	4445.73	5703.17
13. Appropriation to Development Fund along with interest	744.98	449.50	553.85	748.99	1,943.96	2,039.46	2,128.53	2,629.64
14. Appropriation to Capital Fund along with interest	1118.30	1365.95	1842.81	1687.46	1671.43	4086.32	8541.14	11,592.83
15. Appropriation to Railway Safety Fund	..	302.74	..	..	132.46	67.54	..	..
16. Appropriation to Special Railway Safety Fund	..	455.10	602.51	631.57	779.16	748.60	817.66	..
17. Open Line Works - Revenue	35.26	31.53	28.32	33.24	37.56	42.80	51.07	46.79
18. Net Investible Surplus [12 to 17]	4203.97	4629.31	5509.60	5844.09	7457.86	10,827.78	15,984.13	19,972.43

[1] Default in the payment of dividend to the Government of India in 2001 and 2002 was 1823 and 1000 crore rupees (US$ 424 million and US$ 233 million) respectively.

APPENDIX 7

Detailed Financial Results

The data presented here does not reflect the changes in the Railways's accounting method[2] that occurred between 2005 and 2008. Instead for each year, the data reflects the accounting practice prevalent in that year. For comparable data refer Appendix 6, *Statement of Cash and Investible Surplus*.

Note: (1) All amounts in crore rupees; (2) * Includes Safety Surcharge Collection.; (3) **Excludes transferred to Deferred Dividend Liability Account.

Description	2001	2002	2003	2004	2005	2006	2007	2008
1. a. Capital-at-Charge	32,662	36,758	40,709	45,672	48,957	53,062	58,145	63,981
b. Investment from Capital Fund	10,390	10,390	10,390	10,390	10,390	12,816	17,886	24,540
c. Investment in Metropolitan Transport Projects	2999	3280	3593	3944	4264	4476	4728	5130
d. Investment in National Projects	..	..	..	..	1888	2888	3738	4634
Total	46,051	50,428	54,692	60,006	65,499	73,242	84,496	98,285
2. Traffic Receipts								
a. Passengers	10,515	11,196	12,575	13,298	14,113	15,126	17,225	19,844
b. Other Coaching	764	872	988	922	990	1,152	1,718	1,800
c. Goods	23,305	24,845	26,505	27,618	30,778	36,287	41,716	47,435
d. Sundry Other Earnings	703	945	1080	1004	1157	1839	1712	2565
e. Total Earnings (a to d)	35,288	37,858	41,148	42,842	47,038	54,404	62,371	71,645
f. Suspense	-407	-21	-80	63	332	87	361	75
g. Gross Traffic Receipts (e + f)	34,880	37,837	41,068	42,905	47,370	54,491	62,732	71,720

	1130	1520*	1673*	2006*	1676*	1824*	2054*	1557
3. Miscellaneous Receipts	36,010	39,357	42,741	44,911	49,046	56,315	64,786	73,277
4. Total Receipts	27,534	28,703	29,684	30,637	33,389	35,030	37,433	41,033
5. Ordinary Working Expenses	2301	2000	2402	2593	2700	3604	4198	5450
6. Appropriation to Depreciation Reserve Fund	4832	5590	5940	6253	6670	6940	7416	7979
7. Appropriation to Pension Fund	34,667	36,293	38,026	39,483	42,759	45,574	49,047	54,462
8. Total Working Expenses (4 to 6)	272	727	885	950	1,014	2,736	1,286	480
9. Miscellaneous Expenditure	34,939	37,020	38,911	40,433	43,773	48,310	50,333	54,943
10. Total Expenses (7 + 8)	1071	2337	3830	4478	5273	8005	14,453	18,334
11. NET REVENUE (3 – 9)	308**	1,337**	2,665	3,087	2,716	3,005	3,584	4,239
12. Dividend Payment to General Revenues	..	..	50	300	483	663	663	664
13. Payment of Deferred Dividend	308	1337	2715	3387	3199	3668	4247	4903
14. Total Dividend payment (11 + 12)	763	1000	1115	1091	2074	4337	10,206	13,431
15. Excess / Shortfall (11 – 14)	732	449	550	730	1842	1853	1880	2359
16. Appropriation to Development Fund	31	248	..	..	..	2417	8326	11072
17. Appropriation to Capital Fund	..	303	..	..	132	67	..	..
18. Appropriation to Railway Safety Fund	..	..	565	361	100	..	..	..
19. Appropriation to Special Railway Safety Fund	98.3%	96.0%	92.3%	92.1%	91.0%	83.2%	78.7%	75.9%
20. OPERATING RATIO (8/2e)	2.5%	5.0%	7.5%	8.0%	8.9%	15.4%	19.0%	20.7%
21. Ratio of Net Revenue to Capital-at-Charge and investment from Capital Fund (11/(1a+1b))								

² For instances, the criteria to allocate lease charges was revised in 2006. Similarly, other accounting changes were introduced in 2007 and 2008.

Appendix 8

Break-up of Actual Plan Expenditure

Note: All amounts in crore rupees.

Plan heads	2001	2002	2003	2004	2005	2006	2007	2008 provisional
i) Socio economic projects								
New Lines (Construction)	701	860	1315	1493	1690	1991	2488	2667
Gauge Conversion	454	686	812	1164	1121	1242	2136	3022
Metropolitan Transport Projects	263	281	312	351	317	212	253	401
Total (i)	1418	1827	2438	3008	3128	3444	4878	6090
Percentage of total plan expenditure	15	18	21	22	20	18	20	21
ii) Throughput enhancement and other plan expenditure								
Doubling	524	600	578	532	488	687	1202	1671
Traffic Facilities & Computerization	204	240	234	273	373	484	744	944
Rolling Stock & Leased Assets	3639	3056	3479	3784	4541	6623	8060	9506
Track Renewal, Road Safety Works-LCs, ROBs/RUBs	1781	2024	2660	2948	3645	3486	4156	3568
Signalling and Telecommunication Works	350	369	551	690	818	1043	1179	1343
Electrification Projects & Other Electrical Works	395	367	367	268	272	233	452	709
Passengers Amenities	136	169	175	181	223	256	408	668
Other Plan heads	947	1524	927	1709	1935	2583	3923	4181
Total (ii)	7977	8350	8970	10386	12294	15394	20124	22590
Percentage of total plan expenditure	85	82	79	78	80	82	80	79
Total Plan Expenditure (i) and (ii)	9395	10177	11408	13394	15422	18838	25002	28680

Appendix 9A

Excerpts on sustainability, Minister's Budget Speech, Part 1, 26 February 2008.

Vision 2025

Paragraph 62. The financial turnaround of the Railways has been achieved by thinking beyond the beaten path, taking innovative decisions in commercial, operational, and pricing policies and through cross functional cooperation and coordination. For making this magical turnaround durable, we will prepare a Railway Vision 2025 Document within the coming six months which will present new ideas and initiatives in a novel manner. This shall outline our preparedness and strategies for the future. This document will set forth the target for the coming seventeen years in the field of operational performance and quality of service. It will also detail an action plan for achieving the stipulated targets and necessary investment plans thereof. This document will also contain details of customer-centric modern passenger services and various freight schemes to sharpen the competitive edge of Railways. This will have a blueprint of an organization that encourages trans-departmental decision making to take the Railways to unprecedented heights. Route-wise planning would be done to reduce traffic bottlenecks, expand the network and modernize the Railways. The passenger services will be governed by two words 'comfort and convenience'. The buzz word in freight business will be 'commitment and connectivity'. All these efforts will lay a solid foundation for a resurgent Railway. This document will inspire the railway management and its employees to do new experiments, and will be like a guiding light for the future generation (pp. 21–2).

Innovation Promotion Group

Paragraph 63. During the last four years passion for creativity and risk taking has led to the magical turnaround of the Railways. In the twenty-first century, the business scenario is changing fast at the speed of light. It is necessary to make coordinated efforts to face the new challenges and to imbibe new technique and thoughts. Therefore, we have decided to set up a multi-departmental innovation promotion group in the Railway Board. All Railway employees and citizens of the country, will be able to send their innovative suggestions to this group. This group will be provided with appropriate facilities and resources for innovation (p. 22) .

Strategic Business Unit

Paragraph 64. The last four years have seen a rise in Railway's share in transportation of steel, cement, coal etc. To maintain this progress, we have decided to set up a strategic business unit in Railway Board for coal, cement, steel and container traffic to facilitate timely settlement of all problems of our clients through a single window system. This unit will be appropriately empowered for taking full advantage of emerging business opportunities and improving Railways's competitiveness in the market (p. 22).

Information Technology Vision 2012

Paragraph 65. In order to make improvements in operational efficiency, bring transparency in working, and provide better services to the customers, Railways are trying to bring about radical changes in Railway technology systems and processes. For achieving these objectives, attention is being focused on IT applications in three core areas namely freight service management, passenger service management, and general management. For getting maximum benefit in the coming years, the mantra for present and future IT applications would be seamless integration. The Railways's nationwide communication infrastructure will provide the foundation for a common delivery network and platform. Modern technologies like GIS, GPS and RFID will be applied progressively. A centralized information system will not only be useful for the customers but also for the organization as well. The customers will have accurate, fast, and on-line access to information on various subjects. For the customer it would result in superior experience with improvements in overall efficiency, safety of Railway operations, ease of transactions and value added services like infotainment, on-board television, and knowledge kiosks with internet facilities. For the organization, planning and deployment of resources would become much easier with a panoramic view of assets and this would have a multiplier effect on productivity, organizational efficiency, and staff satisfaction. The Vision for IT would be implemented over the next five years (p. 23).

Appendix 9B

Excerpts on sustainability, Minister's Budget Speech, Part 1, 26 February 2007

Railway's New Profile–11th Five Year Plan

Paragraph 35. The year 2007–08 is the first year of the 11th Five Year Plan. Sir: *Ho Izaazat to karun bayan dil apna Sanjon rakkha hai maine rail ka ek sapna* (p. 11).

Paragraph 36. The Railways are targeting a freight loading of 1100 million tons and passenger traffic of 840 crore by the terminal year of the 11th Five Year Plan. In order to make this unprecedented growth a reality, it is absolutely essential that in the next few years the Railway's transport capacity be expanded and doubled, unit cost be brought down by playing the volume game, and customers be provided world-class services (p. 12).

The Quantum Jump in Investment

Paragraph 37. During the Eleventh Five Year Plan, we will invest many times more as compared to earlier plans. There is no readymade investment policy for a vast network like the Indian Railways. The growing demand for transportation can only be met through a harmonious blend of short term and long term policies. Our short term policy of investing in low cost high return projects has been successful in eliminating network bottlenecks and in ensuring effective utilization of rolling stock. Alongside a twin mid-term and long-term investment strategy will be adopted to enhance productivity through, modernization and technological upgradation on the one hand and enhancement of capacity of the network and rolling stock on the other (p. 12).

Paragraph 38. Construction of the Eastern and Western Dedicated Freight Corridors will start in the year 2007–08. These will be completed during the 11th Five Year Plan at a cost of about Rs. 30,000 crores. Even though the Golden Quadrilateral and its diagonals constitute 16 per cent of the rail network, more than 50 per cent of the traffic moves on these routes. As these routes are super-saturated, I have given directions for conducting of pre-feasibility surveys for construction of East-West, East-South, North-South, and South-South corridors. My dream is to construct these corridors in a manner that they develop into efficient and economical trunk routes for speedier, longer, heavy-haul trains. After completion of the freight corridors, the problem of passenger and goods trains running on the same network at different speeds will be solved and most of the level crossings would be converted into ROBs. Where possible ordinary passenger trains running on these routes will be replaced by MEMU and DEMU trains. This will facilitate increase in the speed of passenger trains (p. 12).

Gauge Conversion

Paragraph 39. Sir, on account of partial gauge conversion on various routes of the Railway network, the remaining metre and narrow gauges have become like islands. Cut off from the main network, these lines give the

Railways less than 1 per cent freight traffic whereas they still constitute 20 per cent of the total network. As a result Railways are losing thousands of crores of rupees annually. Even freight traffic has become a losing proposition on these lines. Therefore we will make all efforts to convert the majority of the metre gauge lines to broad gauge during the 11th Five Year Plan. Priority will be given to accord approval for execution of projects which will serve as alternate routes on the network; significant among these are Udaipur-Ahmedabad, Lucknow-Sitapur-Pilibhit-Shahjehanpur, Dhasa-Jetalsar, Jaipur-Sikar-Churu-Jhunjhunu, Ratlam-Khandwa-Akola, Chhindwara-Nainpur, and Ahmedabad-Botad. Projects where the state governments contribute 50 per cent of the total cost would also be given priority in sanction and implementation. Gauge conversion will facilitate integrating the remote and far-flung areas of the country with the national mainstream. Integration with the unigauge network and consequent increase in traffic and reduction in unit cost of these lines will reduce losses being incurred on these lines (p. 13).

Construction of High Speed Passenger Corridor

Paragraph 40. Sir, India is today seen as a rising power in the world. The rapid growth of the economy, rising industrialization and urbanization, and unprecedented growth in intercity travel, has opened infinite possibilities for developing high speed passenger corridors. Hon'ble Prime Minister while laying the foundation of the Western Dedicated Freight Corridor had expressed the hope that the Indian Railways would also develop world class passenger systems. Therefore, we have decided to conduct pre-feasibility studies for construction of high speed passenger corridors, equipped with state of the art signalling and train control systems, for running high speed trains at speeds of 300 to 350 km per hour; one each in the Northern, Western, Southern, and Eastern regions of the country. These trains will cover distances of up to 600 km in two to three hours. All alternatives including Private Public Partnership will be considered for implementation of these corridors. Global warming and changing climatic conditions are a worldwide concern today. These energy efficient and environment friendly systems would go a long way in alleviating these concerns (p. 13).

Suburban Service

Paragraph 41. Suburban services are the lifeline of our nation's commercial capital Mumbai. To mitigate the overcrowding of Mumbai's trains, enhancement in capacity of these services will have to be undertaken during the 11th Five Year Plan. During this plan period MUTP Phase I will be completed. The work on Phase II costing Rs 5000 crore is also proposed to be started. Financing of the Rs 5000 crore MUTP Phase II project will be done with the participation of Railways, state governments and multilateral funding institutions. All out efforts will be made to complete both these phases within the 11th Five Year Plan so that suburban services and long distance trains are completely segregated. This will enhance their capacity by 56 per cent. Ongoing work on suburban services in Kolkata and Chennai will also be completed on priority basis. During the 11th Five Year Plan efforts will be made to introduce air-conditioned class services in suburban trains in Mumbai, Chennai, and Kolkata and escalators at major stations (p. 14).

Rolling Stock Modernization and Capacity Augmentation

Paragraph 42. In view of the demands of growing traffic, along with expansion of network, availability of rolling stock will be increased through effective utilization of available rolling stock, technical upgradation and modernization, and by setting up new production units. During the 11th Five Year Plan production of rolling stock will be doubled as compared with the previous Plan. Capacity of existing rail coach and locomotive production units will be enhanced through expansion of these units. High horse power, energy efficient locomotives with new technology will also be produced. During this plan, production of MEMU, DEMU, and EMU coaches will also be stepped up. One new factory each for rail coaches, diesel locomotives, electric locomotives, and wheels will also be established. The locomotives to be manufactured in these units will be equipped with state of the art technology and will be capable of hauling longer, heavier high axle load trains. The new Rail Coach Factory will produce high capacity, modern and comfortable coaches. Similarly production of 32 tons axle load, higher pay load lower tare weight and track friendly wagons will start for the new Dedicated Freight Corridors (p. 14).

Use of IT in the Railway Services

Paragraph 43. In the Eleventh Five Year Plan, investment in IT projects will be increased to several thousand crore rupees to harness the immense possibilities offered by IT in the interest of Indian Railways. IT applications will be deployed to increase passenger and freight earnings, improve the image of the Railways in the eyes of the customer, reduce operating costs, ensure effective utilization of human and physical resources and to help the top management in arriving at long term policy decisions by developing MIS & LRDSS. A commercial portal will be developed in the next three years for yield management, specially to attract traffic for returning empties and filling up vacant seats. All modules of FOIS including rolling stock maintenance and examination, revenue apportionment, crew management, control charting COIS, etc. will be integrated and implemented in a time-bound manner for completion by 2010. Alongside ERP packages will be implemented in workshops, production units, and selected zonal railways. A common website integrating the more than 50 different websites of Railways will be developed with built in facilities like e-payment and e-tendering. For an integrated approach in IT, CRIS will be entrusted with coordination of all IT applications of the Railways and for development of a comprehensive vision on IT. CRIS will be developed as an autonomous and empowered organization, drawing officers from various Railways services. Indian IT companies have hoisted the national flag all over the world. We invite these companies to take part in various IT projects of the Railways under public private partnership (pp. 14–15).

Railway Electrification

Paragraph 44. The electrified network will be extended over the Golden Quadrilateral and its diagonals, and in all directions from Kashmir to Kanniya Kumari and Guwahati to Amritsar by the end of the 11th Five Year Plan. Electrification of Thiruvananthapuram-Kanniya Kumari, Thrichur-Guruvayur, Tiruchirapalli-Madurai, Barabanki-Gorakhpur-Barauni-Katihar-Guwahati and Jallandhar to Baramullah sections will be completed during the 11th Five Year Plan. In the first phase, electrification of Jalandhar-Jammu, Barabanki-Gorakhpur-Barauni, and Tiruchirapalli-Madurai sections are proposed to be taken up in 2007-08. Similarly, doubling and electrification of Pune-Wadi-Guntkal and electrification of Daitari-Banspani, Haridaspur-Paradeep new lines will be undertaken by Indian Railway's Public Enterprise RVNL in the coming years (p. 15).

Public Private Partnership Schemes

Paragraph 45. Investments at a much larger scale will be required for the above mentioned capacity and expansion network as compared with the provision made in the Tenth Five Year Plan. The funding of this plan of several lakh crores would require multi-source approach based upon deployment of internal resources, market borrowing, public private partnership and budgetary support. The improved financial performance of the Railways will enable a large share of the financing to be met from internal and external budgetary resources. I am not in favour of blind privatization of the Railways nor is PPP a compulsion or fashion for us. We are seeking partnership with the private sector on the terms that are in the interest of Railways and our customers. For example, by leasing out catering and parcel services we have reduced our catering and parcel losses of more than a thousand crores. We have enhanced our capacity by attracting private investments in the wagon investment schemes and siding liberalization schemes etc. Even while retaining the core activity of train operations, we have awarded licenses to private parties for running container trains, which is likely to attract investment of thousands of crores in wagons and construction of terminals over the next few years. We want to have many more such PPP Schemes where one and one make eleven and not two. Public Private Partnership options will be explored with the aim of modernization of metro and mini-metro stations with world class passenger amenities, development of agro retail outlets and supply chains, construction of multi-modal logistic parks, warehouses and budget hotels, expansion of network and increase in production capacity. We have constituted a PPP Cell which will develop the policy framework to provide non-discriminatory level playing field to investors, prepare the bankable documents and set up the procedure for awarding partnerships through open tendering system (pp. 15–16).

Railway Safety

Paragraph 46. Railway safety is our prime concern. I am glad to inform the House that funds for replacement of overaged Railway assets are now provided as soon as the assets become due for replacement. Sir, we have allocated Rs 5500 crore towards Depreciation Reserve Fund for the year 2007–08 as compared to Rs 2100 crore provided in 2001–02. This has had a direct impact on Railway's safety record. Although the gross traffic volume has increased from 724 million train kilometres in 2001–02 to 825 million train kilometres in 2005–06, the number of accidents is expected to be less than 200 in 2006–07 against 473 in the year 2001 (p. 16).

Organizational and Human Resource Development

Paragraph 56. Railways will have to develop a strong management team in which players play, not for themselves, but for the success of the team to tackle complex situations of the competitive market and to fulfill the growing expectations of its customers. A decision in the matter would be taken by December 2007, after evolving a consensus through dialogue with representatives of all Railway services and based upon recommendations of different expert groups constituted so far. The institutions of GMs, DRMs, and CAO (Construction) would be strengthened and developed as profit centre, business unit, and project unit respectively by suitable empowerment (p. 19).

Paragraph 57. Sir, Railway employees and officers require training at periodical intervals in view of rapid changes in the underlying economic and competitive environment. We have, therefore decided that Railway officials will be sent on training to reputed national institutions once every five years and for foreign training every ten years. Officers would have to undergo a mandatory training before promotion to JA, SA, and HA grade (p. 19).

Paragraph 58. The Railway Staff College Building in Vadodara will be renovated in consultation with renowned architects in such a way that its old elegance and splendour is retained even while it is equipped with all modern amenities. Rail Bhavan would be made centrally air-conditioned and a building equipped with modern facilities (pp. 19–20).

Index